Diana,

thanks o

Mr. Hamlin's List

good reads!

Moon

Other Books by Stan Moore

MISTER MOFFAT'S OPUS
A historical novel set in 1917–1929 Colorado. Political maneuvering and financial shenanigans underlie the Moffat Tunnel railroad project. It took five years to build, cost millions, and was the largest construction project in the nation. Mik Mas and Cam Braun are in the thick of the effort.

MISTER MOFFAT'S HILL
A historical novel, 1904–08 Colorado.
Cam Braun and Mik Mas struggle to run trains over the continental divide's Rollins Pass. A diamond mine scheme comes to their attention and fireworks result.

MISTER MOFFAT'S ROAD
A historical novel about David Moffat's railroad from Denver towards Salt Lake City, set in 1902. Mik Mas and friends help Moffat to overcome unforeseen barriers.

OVER THE DAM
Mik Mas uncovers and works to stop eco-vigilantes in today's Summit County, Colorado.
Fiction (overthedam.com)

SEESAW: HOW NOVEMBER '42 SHAPED THE FUTURE
A fresh look at the crux month of WWII.
Nonfiction (seesaw1942.com)

Mister Hamlin's List

Stan Moore

This is a work of fiction. Any resemblance to persons living or dead is entirely coincidental. Some of the towns and establishments described do in fact exist. However, I have taken liberties with their descriptions and of locations and geographical features.

Design by Jack Lenzo

To George S. Henderson, 1874–1904,
the great uncle I never met, and the
many other people who lost their
lives in Colorado's Labor Wars.

Contents

Foreword

This story is about Cripple Creek, 'The Greatest Gold Camp on Earth.' It comprises love, greed, power, friendship, riches, terror, and gold.

The historical occurrences—the public gatherings, the wealth, the explosions and shootings, the bullpens and deportations—happened as described. Some characters and dialogue are devised to add context and realism to the events. Any resemblance to persons living or dead is purely coincidental.

First, my muse, sounding board, and cover artist. Thank you, Kiki. I literally couldn't do it without you.

Others too numerous to mention have contributed; thanks to all. Everyone did their best to help make this a good book and keep me out of trouble. Any errors or misstatements are mine alone.

—Stan Moore

Abby's Maps

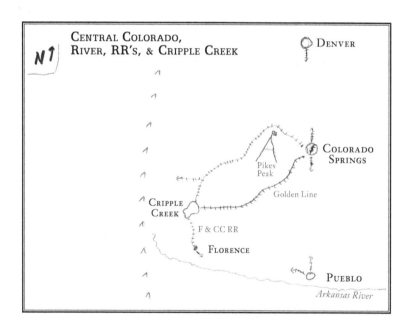

CENTRAL COLORADO,
RIVER, RR'S, & CRIPPLE CREEK

DENVER

Pikes
Peak

COLORADO
SPRINGS

Golden Line

CRIPPLE
CREEK

F & CC RR

FLORENCE

PUEBLO

Arkansas River

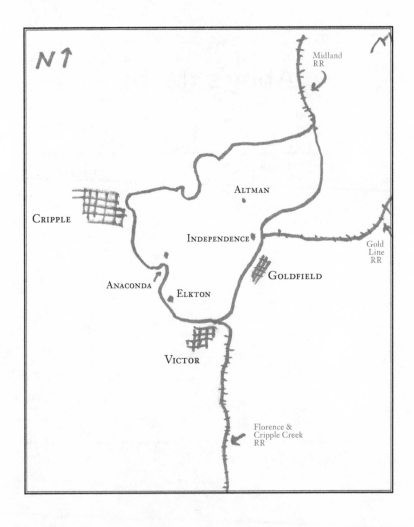

What's What and Who's Who

NOTE, FICTIONAL LOCATIONS ARE DENOTED WITH A DOUBLE
ASTERISK.

Assay Office: A chemistry lab which could assay, or test, the mineral content of ore. Content was reported as ounces of gold per ton of ore.

Bull Pen: Open air corrals in which men were held.

CCMOA: The Cripple Creek Mine Owners Association, also known as 'MOA.' Owners of mines organized to protect their interests. Most were large mines but some smaller mines participated as well.

Cripple Creek: 'The greatest gold camp on earth!' This mining settlement sits in the shadow of Pikes Peak some fifty miles southwest of Colorado Springs. About 50,000 people live and work there in 1900. The main towns are Cripple Creek and Victor. Smaller towns in the camp include Altman, Goldfield, Independence, Elkton, and Anaconda. There are hundreds of mines, some of which generate fabulous riches, running to the millions of dollars. Three railroads serve the area, two from Colorado Springs and one from Florence. The name originated in the late 1880's when a rancher's calf got caught in a fence near a creek and was crippled.

Color: The term used to denote possible valuable mineral presence in rocks or the landscape.

**Double I Mine: A moderately successful mine found and developed by Lon. He named it to honor his and his sister's Irish and Italian heritage.

**Emma May Mine: A mine found by Cook Eisner. It was promising enough to bring in an investor but the vein pinched out and the owners lost their money.

Findley Mine: A profitable mine near the town of Independence.

High grading: Miners and others would steal ore from mines and sell it to shady assay offices. The practice was widespread and authorities turned a blind eye to it.

**Strombo Point: One of many landmarks in the area.

Sun and Moon Mine: An Idaho Springs gold mine.

WFM: The Western Federation of Miners, also known as the WFM, was a militant labor union. It was based in Denver with local chapters in all the regional mining camps.

Who Is Who

FICTIONAL CHARACTERS ARE DENOTED WITH A DOUBLE ASTERISK. Any resemblance of fictional characters to any person living or dead is purely coincidental.

Owens Anderson Owe is a sometime miner and fulltime schemer. He met Juni and Ben on the road and like them he settled in Cripple.

Ed Bell General in the Colorado Militia, sometime Sheriff of Teller County. An associate of Teddy Roosevelt, Bell was a staunch anti union man.

Abby Bosini Abigail Bosini is a school teacher and community activist; has been in town several years. One of her hobbies is maps. She loves to sketch maps of the town, the region, river paths. Anything she can observe or travel through, she draws.

Lon Bosini Dillon Bosini is Abby's brother. He took some training and came to Cripple around the turn of the century; seeks his future.

Bert Carlson As a young man he established himself as the premier hauler. His wagons hauled ore from most all the mines. He owned the Findley Mine among others and died a wealthy man.

Cooker Eisner Cook is a prospector who opened the Emma May Mine. He traded a financial stake from Ben for a one third interest.

Clarence Hamlin His duties as Secretary of the MOA include keeping lists of union and non union miners. Union miners need not apply.

Big Bill Haywood Bill Haywood, a large man, is Executive Secretary of the WFM. A union man through and through, he believes capitalism is evil and needs to be destroyed. He is the driving force behind the WFM.

George Henderson The author's great uncle. A typical young man of the age, he came to Cripple in search of work and wealth.

**Junia Jefferson Juni is another miner fugitive from the crash of '93 looking for work. Came to town with Ben.

**Ben McNall Benjamin Franklin McNall is a miner come to Cripple after the silver crash of '93, looking to find work as a miner.

David Moffat A mine owner, business man, and railroad builder.

Harry Orchard Harry is a union hanger on, schemer, and enforcer. He left Coeur D'Alene Idaho in the late 90's under a cloud and running feud with the Governor. Skills he employs include the use of dynamite.

Spencer Penrose Penrose got started in real estate in Cripple. He went on to own mines and smelters and died a wealthy man.

**Frank Shuler Frank is the president of the local chapter of the Western Federation of Miners.

Winfield Scott Stratton A prospector who for years supported himself as a carpenter. He hit it really big in Cripple. His Independence Mine sold for $10,000,000 before the turn of the century.

Danny Sullivan Postmaster in Cripple Creek. He stepped in and saved Teddy Roosevelt from harm in 1898, and negotiated with union men in 1904.

Jim Warford One of many 'law men' who hired themselves out to provide security for mine owners. He was rough and tumble and was brought in when muscle was called for.

WILL THIS BE THE PLACE? BEN WONDERED AS HE LOOKED OVER the valley.

Shafts of sunlight shone down through clouds, highlighting the mines, shafthouses, homes businesses and mills clinging to the hillsides. The view was reinforced by the streams of smoke dust and steam drifting up from smelters and countless chimneys. The mid day shadows were short and several thousand structures stood, sentinels and makers of gold and wealth. The occasional wafting breeze weaved and distorted the plumes from burning coal and heated ore in smelters. The horizontal smudges overlaid the whole, formed then blown away in moments.

People were everywhere, and the scene was a virtual hive of energy. Crowds on sidewalks, people coming and going to stores saloons and mines, animals pulling and carrying, all gave the impression of an ant colony or a dense bacterial sample under one of the modern new microscopes. Even in the sunlight, electric signs and lights shouted their messages and lit their areas. The wires feeding them were strung crazily on poles and building corners. Trains came and went, adding their toots, chuffing, bells, and calls of 'all aboard!' to the background. Some had heard tell of the 'London Hum,' the sum of economic and physical activity in that city. This city too had its own 'Hum,' one propelled not by manufacturing like London's. This Hum was built on and fueled by taking care of the men who searched for, refined and hauled what they sought: precious metals. Gold.

All of it, the sun, smoke, buildings, people and machines sat high in a treeless bowl of earth, easily twenty miles across and half a mile deep. Pikes Peak loomed to the north and east. Opposite, off to the south and west, the Sangre De Cristo Mountain Range revealed snowy summits a mile higher than the city, their peaks almost three miles above the sea.

Benjamin Franklin McNall hitched his rucksack. It contained all he was willing to carry. Once, not long before, he had owned more but now kept and toted just what he couldn't go without. Actually, he had had to winnow his belongings more than once. This time, he was starting to wonder. Maybe this nomadic life wasn't as fun or glamorous as once thought. And it sure was a lot of work.

Panting from the climb up the lip of the hill, he and his companions stopped. Ben spoke.

"Cripple Creek, Colorado. We're told it is the greatest gold camp on earth. Just looking, it sure seems like quite a place, busy and rich. No doubt in places it is seedy and wholesome and poor and wretched and wonderful. It isn't just one town. Really, there is Cripple Creek, Victor, Goldfield, Bull Hill, and other settlements. All in one district, one gold camp as they say."

McNall bit his tongue, having said enough, and looked at his traveling companions. All three scanned the sights, solemn as junior undertakers, and wondered how they would fare in a year or two.

Ben put words to the thoughts. "All I can say is, I hope this town treats me better than Aspen did."

Junia Jefferson, Juni to those he liked, spoke.

"Yah. Me too. Not Aspen for me. They get too much snow there every winter. Me, I'm coming from Gilman. The town on edge, north of Leadville, hanging on to the side of the mountain, God knows how. After that I had a stopover in Creede. That lovely town—what's the line, 'It is day all day in the daytime and there is no night in Creede'? Well, there is puh-lenty of night time there now. Silver towns all around are going dark, or at least slowing way down."

The memory of the poem made him wistful and he stopped a moment, still looking at Cripple, then went on about Creede.

"It used to be not far off the wagon route from Alamosa to Silverton, but no more. Not with the train running south out of Alamosa over to Durango then north to Silverton. Heck, that way is so much easier than going over the mountain passes on the old straight wagon road route past the big C. I can't believe we used to think that was convenient."

A rumble of thunder way off, but no flash of lightning, made him pause, look up and around. But it was so far away he wasn't worried.

"Anyway, actually...where was I? Oh yeah, actually I was running out of Colorado mining towns to beg work in. So I even thought of going out to Carson City in Nevada. The Comstock! Good thing I heard before I left that the mines even there were was not hiring. They are laying off too."

All this was from Ben's new friend Juni. Well, not new-new. They had met on the road several weeks back. First they traveled parallel, then got to talking, then became friends.

Junia Jefferson, quite a name. Most hearing it would likely expect a slight, maybe even a female person. It was after all kind of an indefinite name, one Ben thought no kid should be saddled with. He learned later the man was named after some great uncle who had been a well known Viking scholar. Whatever, Ben thought. Juni had some swarthy background, some Mediterranean or African blood and how an uncle came to study Vikings was no doubt a story in itself.

In any case, not at all was Juni of the womanly persuasion. The man was six feet two inches tall and carried at least two hundred pounds. Not one of those pounds was flab. He was the image of muscular.

Growing up, he started small and got his growth late. So he had to learn to fight. At age fourteen or so he shot up and put on weight. As that happened, of course, boys and other bullies no longer wanted to scrap with him. He learned not to take advantage of his size but would use it if need be, for himself or to help another. As a man he was careful to take people as they were. As a boy he had to put up with people assuming things about him. He hated that, he wanted to take people right. Juni tried never to assume anything about a person, rather to observe and ask and find out. He didn't want to inflict that behavior on others.

The third miner Ben and Juni walked with put in his two cents. He too spoke of another town hemorrhaging money and miners. "It couldn't be any worse here than it was in Leadville. One day I was working and getting on just fine, the next the bottom fell out."

Ben grinned ruefully. "That happened to a lot of us, going around looking for work. Us miners, of course, since all of a sudden silver wasn't worth the effort to dig it out of the ground. But the hardware merchants and saloonkeepers and their girls and the grocers and teamsters all fell out of work too. I'm still not sure how that happened so darn quick. That was just silver, but gold! Gold is different. Gold is, well, golden—the very meaning of value and wealth. People have had gold fever since the Egyptians and before! And Cripple Creek has lots of it. Most of what is here hasn't even been discovered yet, I'd say."

Juni looked around; the three were not alone. There were solitary travelers and several groups of men heading down into town. Some had a modicum of equipment—shovel, pick or hammer and some owned only the clothes on their back.

"Yah, lots of gold. But there are lots of miners too. Lots."

"You got that right, Juni. Like I said, most of us are silver miners. Or maybe it is, we were silver miners. Now we're just out of work silver men. Not long ago we couldn't dig enough of the stuff. Now, we're looking for a new job. Heck, we might even have to leave mining, which I don't want to do. You are so right, there are many out there like you and me."

Ben again scanned the view in front of them, the town and mines and railroads spread out like slumgullion stew in a great big shallow bowl.

Juni followed his thoughts. "And all of us out of work silver dopes are needing jobs. Mining jobs. There are more miners than jobs. That is probably true even here in Cripple Creek. Of course that means with the competition for openings, many will be willing to work at any rate. You watch, that will become a problem, men working for cheap, maybe next to nothing."

The man who came from Leadville went by Owens Anderson.

"Call me Owe," he said, when they introduced themselves. He nodded at Juni, agreeing that wages would be pushed down with the glut of rock knockers. He had something to add.

Anderson was fire to Juni's ice. He brought to mind a praying mantis, a very talkative and energetic praying mantis. Owe had a small head—unusually small. In many it would have been the one thing people remembered. But Anderson's head was literally overshadowed by gawky, spasmodic arms and long legs taking strides, either in a hurry to go somewhere or pacing in circles. If Owens Anderson was awake, his arms and legs were in motion. He didn't always talk but when he did, his tongue was as active as his limbs.

"And the mine owners will take advantage of that, you better know it, they'll use that to the fullest they can, like dumping a barrel of water into a pail. Those owners will try to drive wages down, wages down and hours up, you bet. Us working miners better be ready to resist 'cause those greedy sons of bitches don't care they just want money and don't give a rip about who they use up to get it. And a bunch of new yay-hoos coming to town looking for work will help them and hurt us miners. Say did you guys hear about what our navy did over in the Philippine Islands, how our ships destroyed the Spanish Fleet in Manila Bay? And now we have landed troops to take it away from the Spanish which makes me ask, why do we want a bunch of islands ten thousand miles away anyway? Gosh maybe there is gold over there—that might be a place a miner could find work."

His conversation ran down and stopped. It reminded Juni of how a top would spin fast for a while then lose momentum and finally stop. Owe was apparently now thinking about being a miner in the tropics.

Ben snorted. "Hey, who cares about gold in the south seas. Mine owners right here in Colorado are doing alright. Better than alright, most of them. These guys are making money hand over fist with the wages they pay now. It doesn't make sense for them to upset things. Do you really think they'll tighten the screws now that more miners are hungry and available? I mean, really, if things are working well for you, there's no reason to change them around. Is there?"

Owens was transported back to Colorado. He and Juni exchanged a wry glance. Juni spoke; Owens' hands and face twitched but he didn't interrupt.

"Not only yes, but hell yes. Money, making money, is the name of the game. No way would one of those owners leave a dime on the table. Owe is right. They will turn the screws, again and again. Until men refuse to work. And I don't see that happening soon." Juni's expression turned grave.

"We'll each of us have to fight for a job. If one of us gets one, we may have to take it at cut wages. Even so most will be glad to have that. Remember, there's hundreds out there like us." Again he looked around at others walking towards the gold camp. He imagined they were likely talking about job prospects.

That seemed a very bleak prediction to Ben and he didn't really agree. Wages would remain stable if only because good miners were relatively rare. There were plenty of men willing to load muck, pieces of ore, into a car. That, a trained monkey could do. But not everyone could read the vein of ore and how it lay in the surrounding rock. Nor could any joe off the street drill and blast to bring the ore down but not the roof or dross. Now, that set of skills was not so common. It took years of experience based on solid smarts to be able to do that.

McNall was a firm believer that people were able and aware and usually decided well for themselves. Not always, but usually. Anyway, he chose not to make a point of disagreeing. Maybe he was wrong. It wouldn't hurt to be careful in any case.

"If that's the case, if the owners will try to push wages down and hours up, we need to stick together. Let's make a pact, you guys. Owens, Juni. Let's agree, we help and look out for each other. Us three. Alright? We need to have the other guy's back here in..."

Owens interrupted. He was on a toot, back in Colorado from the south seas, harping on labor conditions, one of his favorite issues.

"I'm with Juni. Heck yes, you can bet the owners will try to turn us against each other, not just you two and me, but every man jack. Those mine owners will do their best to make us compete. They'll try to encourage us to cut each other's throat, you just watch. The rich owners and bankers and mill owners and managers don't give a damn about us, we're just cogs in their machinery."

He warmed to the subject, waving his arms and shifting his weight, turning his head from side to side. Ben was reminded of some news accounts he had read. The artist's sketches included really made the story. The tale was of the Moros, Muslim natives, running amok in the Philippine Islands, shouting and waving a big knife as they waded in to kill for Allah. Ben was kind of glad his friend was not armed. If so he might have run off looking for a mine superintendent to skewer. Anderson continued even as Ben stifled a grin at the thought.

"Hell, we may as well be identical and replaceable machine parts as far as they are concerned, far as they even think of us miners. No patience if one of us gets injured or uppity about our pay. If that happens they just shove us out and bring in some other man, the next man, and if we're living in company houses they shove our family out as well. Cold. If the new man causes problems or a slow-down, same thing, out the door with him. At least it is out the door, not down the shaft. Anyway, those damn rich owners make me think of a shark circling a group of baby seals, or a pack of coyotes taking rabbits, or Indians on horseback riding around a small bunch of cavalry, isolated and about to get wiped out. Hell and damn, maybe those owners care even less than that."

Having said his piece, Owens now stood silent and pretty much still. He looked at his friends, awaiting agreement. Ben and Juni shared a glance and a shrug. Then Ben spoke.

"That sure was intense, Owe. Do you know something we don't?"

"I was in Idaho and then Telluride before Leadville. I've seen what mine owners are willing to do, what they'll do to protect their bulging wallets. I saw shenanigans and beatings and troops called out—troops to beat on American citizens! I saw that kind of thing in each of those mining camps and have heard tales about that happening in other camps as well. The owners have the governor in their pocket, or at least on their side, and they can use the Militia like their private cops. So I have little faith in them."

This was not news to Juni. He laughed.

"Really? I am surprised to hear you are skeptical of the men who run things. You mean the little guy has to drop and cover his head when the big money boys decide to do something? Gosh, I had no idea about all of that."

Owens scowled, then chuckled. "Well, I get worked up sometimes. Yeah, really, I know what they will do to keep the ore coming and the costs down and their profits up. Don't get me wrong, there is nothing wrong with making a profit, and I don't like big money men but they shouldn't have to pay more than they make, no one should. But if a big money man is in fact making a profit, he ought to share some of it. Seems to me the man actually pulling the ore out of the earth ought to get a share of that profit, that's all I'm saying."

Ben figured he had heard enough circular arguments. And he knew there was no one answer and such talk would go on for now and for the future. He turned back to the view.

"Enough political talk. I don't care for poor me or for doom and gloom. I mean, we're Americans! We can do what we want, say what we want, go where we want, meet with whoever we want. Lots of folks can't do that kind of thing."

"Yah. My great uncle got in trouble when he talked about how vicious the Vikings were."

"Ha. My father left home and came over to the US to avoid getting drafted into the Imperial Army. Something about fighting for or against Napoleon or Bismarck or someone, I never got the story straight, but here I am and you are right Ben we have good things here. I just want more, Ben, and I want the owners to have less."

"Well, gentlemen, be that as it may. Think about it: Here it is, just a couple of years until 1900. We have a new century, and it will be a great one! New inventions, new ideas, new gold camp! New mines to be discovered! We'll get jobs, and there are plenty here. Heck, I intend to go look and maybe I'll stake a claim or two myself."

Juni: "You are right Ben. We have lots of opportunities. We are here now, and we may as well make the best of it. I agree, let's keep in touch and look out for each other."

Now Ben was on a toot. "Cripple Creek sure is a going jenny of a gold camp, isn't it? Greatest Place on Earth! So says the Chamber of Commerce! Electric lights, people and houses everywhere, mines, smelters, railroads...You can feel the excitement. There is money to be made here, guys, I am sure of it. And better to be here than mining for gold up in the Yukon, fighting winter and grizzly bears." He glanced at Owe. "Or in the tropics swatting flies and dodging Moros running amok. Hey, this place, Cripple Creek, has already coined a number of millionaires, and lots of others are comfortable if not filthy rich. And we're here!"

Juni deadpanned. "And I'm sure happy to be here prospecting. Think of those poor guys who are fighting wars. Like Owe said, our government has troops in the Philippines and Cuba, why we maybe aren't sure. And speaking of war, another one is cooking between Japan and Russia. They're fighting over in China and Siberia, godforsaken cold country for sure. And there's one going on in South Africa too. The British Army and the Dutch Boer settlers sure are going at it. I read or heard somewhere that ranches in this part of the country are making money hand over fist selling horses to the Royal Army, for cavalry and so forth. I guess if I can't get a mining job I'll go wrangle horses somewhere."

Owens twitched and grinned. "Hell and damn, no horses. I bet there is so much rich ore here that the owners can't keep track of it all. High grade gravel probably litters the roads."

"Yah, you're right. Some of the ore taken out of the mines is rich, very rich. So rich that armed guards accompany every wagon load of ore. Can you imagine? Carrying a rifle or shotgun to make sure some kid or hungry miner doesn't take a rock?"

"Only in America, Juni."

He went on. "Most of the output isn't like that. And no, Owe, ore isn't used to pave the streets. Most of what is taken out is good paying quality ore. Average ore is good, good enough that there are shakedowns after every shift. I hear that in some mines, every man coming out of the hole has to prove he isn't taking anything he shouldn't out of the mine."

Owens grinned. "Why Juni, are you talking about high grading, I'm shocked that you would even know about it, much less think of such a thing."

The big man shrugged. "The mine owner might claim right to all the rocks in his place. But you notice, the big rich men might go down the lift once every week or two. To take a look around for a few minutes. They never go down the hole to actually swing a pick. Or muck. Or drill or blast. They are too busy counting their dollars to do the actual work to loosen the rich ore and bring it up to be smelted and refined."

"No, the fat cats would rather pay you and me near starvation wages to do that. Hell and damn, neither David Moffat nor William Jackson Palmer even know which is the business end of a shovel."

Both of these men were railroad entrepreneurs and owned a variety of business interests outside of mining. They were Coloradans, westerners, and kept their money and investments in the region. They were in the minority that way.

Ben smiled coolly. "Now Owens, those two are not bad men, and I am sure they do know about picks and shovels. Really, of all the fat cats those two are tops. They're owners and rich and not perfect, I'll give you that. But at least they care about what happens in Colorado and to us Coloradans. Most of the big money men don't give a rip about anybody in this state, rancher or merchant or housewife or miner. All they want is to take their ore out and send the gold east." "Yah, you say Moffat. I say Shmoffat. I don't care about any of those guys. What I do care about is, a miner ought to have the chance to keep a little bit of the ore he dug out. If he finds a way to get it out of the mine and sell it, so much the better. The precious mine owner will never miss a few ounces of rock."

With that Juni looked at his companions. "I tell you this, and I tell it as a friend. If I have the chance to do that, get ore out for myself, I likely will. I don't care about what you do; that's your business. And know this: If either of you ever repeats what I just said, I'll deny it. And then I'll come talk to you, real personal like." With that, he made a warrior face, made two fists and jabbed the air.

"Listen to us. We were just talking about looking out for each other, having the other guy's back. And here we are now talking about stealing for ourselves and to hell with everyone else." Ben glanced between the two, then smiled a peace offering.

"Gold does funny things, doesn't it? The prospect of finding lots of that warm soft yellow stuff can hypnotize you, make you act crazy."

Owens shrugged, adjusted his rucksack, and stepped off down towards town. "Well, I for one am ready to take that on. I want to go make big money and have people after me to be friends and turn away women and live in a mansion!"

Ben and Juni laughed as they too headed down to grapple with their futures.

II

Coming east down the hill, Ben looked over at Pikes Peak. Even though the sun was out, it loomed majestic and shadowy in places. Its summit was bare, several thousand feet above the tree line. Ridges, boulders, the peak itself made the light play and change. He could have watched it for hours. Pikes was to the northeast. He looked east then south over lower timber covered mountains and ridges. South, the terrain dropped away and he could see a creek's valley. Up it came a train, and he figured that was the Florence and Cripple Creek Railroad. That was the first train line to reach Cripple, several years back.

Looking on around to the southwest he saw many big peaks, like Pikes craggy with some snow and light playing over them. Behind him, from where he came to the west, was a series of high valleys and plains. Beyond them he knew ran the Arkansas River, from near Leadville south then east. More high valleys and plains lay to his north and west. He looked on around the compass circle. From the north came another train. This was part of the Midland Railroad group he knew. It connected to the line that company ran west from Colorado Springs.

Cripple Creek, the crown jewel, sat in the middle of all this, near the bottom of a huge basin. Ben read up on geology and geography as he could. The theory was that the basin was in fact a huge, old volcanic crater of some sort. The shifting lava, the heating and cooling, had somehow brought the ores of gold and other minerals to within reach.

This puddle of gold, so to speak, brought to mind all the hoohaw about gold and its uses and effects. There was, had been for years, lots of talk about gold and precious metals and the economy and jobs. This was in all the newspapers, at the barber shop, in church, around the dinner table and among friends working in the factories, mines, and fields. He mused on where he fit into all of that.

Ben thought back to 1890, not even ten years back. He had seen and done a lot since then.

He managed to finish six years of schooling. There were gaps but he could read, do sums and work numbers, and knew something of history and geography. After that school term, at age fifteen he stood and surveyed the farm he grew up on and was expected to stay with. Farming was well and good, and he knew his brothers loved the land and the life. The farm and the family was in good hands. Ben wanted more than seasonal rhythms and dawn to dark labor. The horizon called, the western horizon.

He soon left Ohio. With mixed feelings he waved to his mother, father, and siblings as he walked away. He walked only a few days to the tracks. The plan was to jump a freight train headed towards the sunset. He had very little money and in no way could buy a ticket. Being a non-paying guest of the railroad, it was not a quick trip. He had to make many hops on and off. Sometimes it was to avoid being caught, often to find a few day's work to gather funds. And to scrounge something to eat.

A few nights of the journey were spent behind bars. The local law men didn't want him to stay more than one night. They wanted him gone, preferably in the next county. So they simply moved him on. He got something to eat those nights, beans bread and coffee or the like. But that did the job. Along the way, he had more than a few run-ins with yard dicks. The railroads paid these men to keep their cars and yards orderly and free of hangers on, men like Ben. Private security men didn't observe the nicety of 'innocent until proven guilty' nor were they gentle. The first time was truly a surprise; it made him wary and he did his best to avoid contact. Of course the yard dicks were on to most of the tricks free riders tried, so there were more confrontations. Such is the life.

It took a while. Ben had adventures aplenty and garnered a few bruises. The experience gave him sympathy for those at the bottom of the heap. Men along the way helped him and he gained appreciation for how working together could make things easier. He never forgot that.

At last he made it to Colorado. Denver was a big, growing city and he had little trouble landing a job. With his first pay in hand, he actually rented a room and settled in temporarily. Still restless, he intended to move on when he could. It took some time to build savings, and he lived frugally. His wages as a store clerk weren't rich. With careful management more went to savings than to expenses. He was ready to move on before he had the cash. He wanted to get out of the city: there were too many people.

Almost a year later he felt he had the funds and knowledge for the next step. He intended to work at a skilled job, not as a clerk. He bought a cheap seat on the train to Aspen. Not as big as Denver, Aspen was not small. Its population of five thousand or so was more people than lived and farmed in his native Ohio county.

Following the plan, Ben took a job in the mines. First he simply did errands for the old hands, trying to make himself useful, and learn his way around. Then he got to move ore rocks by hand and with a shovel, from a heap into an ore car. Mucking, it was called. The task reminded him of many on the farm, especially cleaning out horse stables. It definitely called for a 'strong back weak mind' but he was happy to do it. He could watch and listen, learning how things flowed and who did what. Muckers were the unskilled, the bottom of the heap.

His experiences on the road came back to him. Day three as a mucker. He leaned over to pick up a rock.

"Hell, boy, what are you doing?" An old man, must have been at least twenty five or maybe even thirty, growled at him.

"Getting this piece of ore, why?"

"You'll spring your back doing it that way. Do you want to be stove up and unable to move around in a year? If you do, just keep doing that."

"Well then, what should I do, how should I do it?"

"Don't bend over and lift with your back. If you have to lift something, squat, keeping your back straight, and lift the weight with your legs. Stronger and you'll last longer. Same with the shovel, bend the knees and lift with your legs as much as you can. Save yourself, young man."

"Thanks."

"Us miners have to stick together. The owner will use you and me up like matchsticks if we let them."

They went back to filling the ore car. Ben tried and it did feel better to use the legs.

"You're right, thanks again. Do you really think the owner doesn't care about us?"

The man laughed. "You really are new to the mines aren't you? I tell you, boy, we're just tools to be used and discarded, far as they care. We need to talk to one another and help each other out."

And so it went. It wasn't all that long before he was able to work his way into better jobs.

The labor was hard; he loved it. It was all good, the exhaustion of a full day's work. Ben really loved learning the trade. He built the skills and knowledge to survive and thrive in a dangerous and demanding but glamorous job.

His brothers and friends in Ohio would be astonished. He learned where to drill to safely bring down a heap of ore and how to recognize and shore up a dodgy spot. He looked forward to work every day, or night, depending on his shift. There was the satisfaction of creating wealth, pure silver wealth, literally making it from rocks, and the camaraderie. Miners were good friends and dependable coworkers. He imagined that soldiers felt a similar fellowship. Maybe in the mines they weren't dodging bullets. Even so, every day down the hole each and every one of them put his life in the hands of his mates. It made for strong bonds and loyalties.

Miners were discouraged openly and subtly from talking to each other. Sure, comparing ideas to decide where to drill or put a support beam was acceptable. But discussing work conditions was not. Comparing your pay and hours with men at other mines was not openly done. Not if you wanted to keep your job or be able to get another. Owners talked and Ben was pretty sure there were lists of miners that owners didn't want around. Being too loud and obvious to discuss conditions landed men on such lists.

After discreet talks with trusted workmates and some reading in his spare time, the light came on.

Modern America was grappling with a basic truth of the mining camp, the factory, the paper mill, the steel works. The miner (or other worker) wouldn't have a job but for the owners' capital. That said, the owner would have no ore to refine but for the miners' labor. Same for the shirt factory, mill, refinery, what have you. So why were those two groups so often at war, labor and capital? Why didn't they work together to maximize good for all of them?

Ben didn't seek enemies. For him the foremen, superintendents and owners were not the foe. Not exactly anyway. That said, his friends came from among the miners and a few other tradesmen. He fell in with the men who did the physical work. Aspen was roaring and for several years this went on. The town and its residents got used to steady work, reasonable hours and pay. All in all everyone seemed to be getting along fairly well.

One day he showed up for his shift. He and his buddies couldn't get in. The gate was locked; no one seemed to be around or coming off the earlier shift. The crews milled around, confused and concerned.

Ben and his mates guessed and speculated.

"I've never seen this before. I wonder what is going on. It can't be good. Can it?"

"I wonder where the shift coming off work is. I hope the mine is safe and no one is hurt and that they're alright."

A foreman came from the shaft building and stood. The slatted shadows of the gate fell on his face and he hesitated, gate between him and them. He looked haggard.

"Bad news, men. The mine has closed. You can pick up your final pay tomorrow starting at 8:00. We will hold back any amounts you owe the company."

Someone in the crowd yelled. "Tomorrow? Why tomorrow?"

Another called out, "Closed? Why? We're moving ore like crazy! Fast as we can. What else does the owner want? How can we get him to reopen? Why is it closed??"

The foreman reached for and held up a newspaper. The headline screamed, 'Sherman Silver Act Repealed!'

"This is why. The government has put the kibosh on. It isn't buying any more silver. They were buying millions of ounces each year. At a guaranteed price. That is what kept us open. They've stopped and all of a sudden the price has dropped through the damned floor. We can't keep the mine open at the going price, no one can I imagine."

He looked around at the men, making eye contact with each.

"If it makes you feel better, I'm out of a job too. Me and the other foremen will be on the street as soon as we get you men paid off."

"But, my wife and kids. My house. My job. What will I do? What will we do?"

The foreman shrugged. "I do not know, my friend. I do not know."

This was stunning news. Aspenites were gobsmacked. People on sidewalks and in the street were worried, shocked. Many walked as if in a trance. There were already 'closed, out of business' signs in a few store windows.

This was true in every mining town across the west. Silver was suddenly worthless, not worth mining like the foreman said. Silver towns were echoing with despair.

On the other hand, gold was holding value. Jobs of gold miners were fairly safe. But many thousands of silver men were suddenly unemployed. Of course, so were the workers that supported them—bakers, teachers, blacksmiths, madams, dance girls, storekeepers, tavern keepers, all were soon adrift.

Ben had it better than many. He was young, single, and could easily move around to find work. With no family to support or house to pay for he was free. First he tried to get on at a nearby gold mine, but of course so did every other miner in the county. With nothing to lose, he even tried at some of the struggling silver operations. Before long those unfortunates put up signs, 'not hiring, no jobs.'

Next was a trip around Colorado. He visited Leadville, Breckenridge, Fairplay, even Nederland and Central City. He tried the mills in Denver, Durango, and Colorado City. No luck. Hard times came with the government's abandoning silver. There were darn few jobs and many job seekers.

Thus he gravitated towards the Pikes Peak region and its mines. This was practically the only viable mine area left after his visits to the others. The greatest gold camp on earth, they said. Hah! Every gold camp made that claim. He was hopeful, though, as he walked down a ridge towards the town. His excitement and interest grew.

After all, this was THE gold camp, Cripple Creek. It was an exciting, growing, crazy gold camp. Cripple was the main town but there was also Victor, Goldfield, Elkton, Altman and others. Hell, it had been going and growing for years and still people were filing claims and striking it rich! Word had it that there were jobs for any man willing to work. Apparently there was no rumor to that truth!

"BEN." JUNI NUDGED HIM WITH AN ELBOW. "BEN, HAVE YOU thought about living? I mean, a boarding house or a room with a family or what? I guess there are some choices for that now. Early on, not so. A man I met in another camp told me that once housing was scarce and expensive. It was so tight that men paid to sleep on pool tables, or on the floor of a tavern. Can you imagine? But like I say I guess that is a thing of the past."

"Yeah, I heard stories like that too. Happened in many new camps, I guess. Looking out at the town, I have to think that is better now. With several railroads and everything else, it is a real town now. It has places, civilized places, for men to live. Hear tell that some mines have housing for their workers, right there on site. Convenient but you're locked in, if you know what I mean.

"Yah, you have to do what the foreman tells you, even after hours."

Ben nodded. "I like to go to town after work, know what I mean? And, I don't want to be beholden to the owner for a roof over my head. Or anything else. The thing is, he has more control over me than I'd like anyway. Even without having me use his shelter!"

Owens overheard and jumped in. Almost literally.

"Yes, I agree with you guys, and if you do that, live there, like you say they have their hooks in. They deduct rent from your pay, and the rates they charge aren't, well, I hear you have to really back the owners into a corner to get them to tell you the actual rate per day and you know if it is an owner it will be expensive. Hell and damn, it could maybe cost as much as a dollar per day, you know, more than you would pay at a regular boarding house although I know of some miners who 'board' in the sporting clubs, because some of the madams want a man around to keep the peace so to speak but I wouldn't do that because it might be tempting to take your pay in benefits if…"

"Jeez, Owe, take it easy." Juni's mouth smiled but his tone of voice said 'shut the hell up.'

"We get your point, man, we get it. Me, I want to find a house, a regular boarding house. Run by some nice respectable woman, maybe a miner's wife or widow, who takes in just a few roomers, maybe three or four men. They could become like brothers maybe. What I really don't want is big. Don't care for a big dormitory type place. That reminds me of the church orphanage I spent some time in as a youngster. Not something I want to relive, yah that's for sure."

He smiled, glad to change the subject.

"Of course, I'd really like to meet a woman, marry and start a family. But that will be hard to do here. If this is like the other places I've been, nice unattached women are rare."

Owens snickered. "Why Juni. There are plenty of women here, I'll bet. Who needs them unattached?"

Juni scowled. "Maybe you don't. I do."

Ben chuckled.

"Juni, you're probably right about not too many nice unattached females. But it won't always be so. It usually goes, that with time things settle down. There will be schools and churches and families. More like a city and less like a wild west outpost. Soon meeting a good woman will be possible. But for now, not likely. But we're getting ahead of ourselves, talking about pairing up already. Being new in town we will likely have to take what housing we get, at least at first."

Ben in fact had an idea or two but wasn't ready to share with his buddies. He changed the subject.

"I hear the going pay here is three dollars for a nine hour day. Of course that is nine hours in the mine with half an hour or more of dead time before and after. You know, when you have to go set up the tools, get the equipment ready and so on before hand, and take it down or put it to bed after the shift."

"Right now a friend told me that the going rate is three for eight hours, not a nine hour day. Of course most owners would like it at nine hours but for now, it is usually eight."

Owens frowned. "But you still have the dead time work before and after. That just seems wrong. It is work, isn't it? It should be part of the shift. Men go into the mine to prepare to pull ore out for the owner, right, so it seems to me that amount of extra time, all of it, ought to count as part of the shift and be paid for and that will become an issue, a workers' battle cry, you wait and see."

"Owens, do you ever take a breath? You talk so long and fast it is hard to follow sometime. But I heard something about battle cry, didn't I? Battle cry?"

Owe nodded, said nothing.

"Jeez," Juni groaned, "you don't really want battle, do you? I've not fought, but my father was wounded and finally died from getting shot. Was in the Cavalry fighting Indians. Bad stuff, you don't want battle, Mr. Anderson. Really, all most of us want to do is go into the mine at a fair wage. We just want to make a living."

Owens again nodded, ready and willing to fire his verbal bullets again.

"Don't we all, Juni, don't we all. One way we can guarantee a fair wage is the union, the Western Federation of Miners, we need to join the union. The WFM has some members here and they want to improve our lives and the conditions we work in and they have stores for us so we don't have to go to the company owned stores and the WFM stores take cash which the many of the company stores don't they take only scrip which forces the miner to stay in hock."

Ben jumped in as his friend took breath.

"Yeah, that is something. I never have liked the company store routine. It is good for the company, another profit center for the owner, like company owned housing. Is it good for the worker? Not so much."

Owens tried to jump in again, this time starting in about the clubs.

"And sporting clubs there are lots of clubs and places to meet others and don't forget the girls and the social clubs I can't wait..."

Juni ignored him as Ben went on.

"There are plenty of stores and enough different mines that a man has the choice of more than just one employer. Like I said, freedom of choice. With all the options, a miner can quit and go to work somewhere else. So the mine owners can't really force their workers to use the company store and pay only in company scrip."

He frowned. "That method, company stores and scrip, is flat out indentured servitude, almost slavery. Didn't the Civil War get fought to stop slavery? Hell, stores and scrip are just a way for mine owners to control their workers. It sure as heck is not a good way for a man to make an honest, independent living."

"Company stores often do have a good selection of goods, I have to say."

"Maybe so, Juni, but at what cost? Who wants to go to a store you have to buy from?"

Conversation lagged, and Ben thought back on that headline, the one the foreman held up in Aspen. It shouted out the repeal of the silver act.

After it happened, as he hit the road, he tried to read up, learn what he could. He learned many of the reasons for the Panic of 1893 as it came to be called. The repeal of the Sherman Act was just part of the picture. It was one of those times when a bunch of activities and decisions came together in a perfect maelstrom of misery and destruction.

When thousands of jobs evaporate and people become poor overnight, fear is rampant, and understandable. And in this case, it wasn't just miners who lost jobs. Banks failed, railroads went bankrupt, hunger stalked the land. For many Americans, maybe most of them, it was a tough time.

He read papers and books as much as he could while he bounced from town to town looking for work. Some of it was gobbledygook about international finance. That part he really didn't understand, but a lot of background and reasons for the problems, he got.

Farmers were doing well and had bumper crops. The surplus meant falling corn prices and failing farms. Railroads which had extended their rails to rural areas lost business which pushed many of them into bankruptcy. That had ripple effects through the service and manufacturing and industries and the towns where they pulled back. The ripples grew and widened.

To add to the chaos, some people speculated. These were not folks who worked with their hands and backs. The buying and selling was to take advantage of chinks in the price of silver and gold. As with much financial speculation, some made money but many more lost it. It didn't help that the government yanked support for the price of silver in the midst of all of this. The President supported the legislation to drop the guarantee for the price of silver. He wanted to stop buying tons every month.

The spreading misery and poverty made for trouble, social and political outcry, everywhere. Farmers in the heartland formed alliances with workers in the cities. These were aimed against the elite and rich, that is the bankers, owners and managers.

Ben read of an Ohio businessman named Coxey, Jacob Coxey. Actually Ben knew the family back on the farm growing up. Coxey spearheaded a protest march of unemployed who went east. Their intent was to protest the handling of the downturn and to petition the Government to create jobs. The prospect of hordes of unemployed men and their hangers on sent shivers through Washington. No public works or relief came from the effort.

There were other actions and plenty of reasons for the collapse. Conditions were tough all over the country. Soup kitchens, tent cities, and other emergency measures were being taken everywhere, not just mining areas. People hurt border to border, sea to sea. Many business owners and farmers simply walked away.

Some cities were trying new ways to help. Detroit opened up garden space for people. They were called 'Pingree's Potato Patch,' after the mayor, Hazen S. Pingree. Ben didn't know it but in forty or so years, he would live to see 'Hoovervilles.' These were tent cities all over the country and the occupants were angry, hungry desperate homeless folks caught up in another economic tailspin. Just as angry and desperately perplexed as the people in 1893 and 94.

As far as Ben was concerned, the upshot for miners from the Panic of 93 was simple: the price of silver plummeted. Mines depending on selling to the government closed. Many thousands of hard rock men were out of work. Suddenly lots of citizens of the silver towns were coming to cities and to goldmine camps, looking for work at almost any wage. The mine owners loved it!

Owens was thinking of similar things, and he might as well have been reading Ben's mind.

"These hard times sure are bad, no? The damned oily rich businessmen and their hired hands the bankers and they have the governor in their pocket and all they want is more money and the working man be damned, I say we need to organize. Without us providing labor for them they are helpless, helpless as a month old baby and I think it is time we rubbed their noses in it and"

Ben interrupted. "And if you don't work how are you going to feed yourself? And how will the worker feed his family while we are 'rubbing their noses'? I agree, this Panic was brought on by greed, greedy rich men and misguided government policies. And not only rich folks, but also by others. Workers and farmers who didn't save. People who ought to know better, but who speculated on gold and silver, who thought prices would stay high forever and who didn't plan.

He shrugged. "Hell, I don't know. I agree we need to change some things, and I want to help, but we'll get nowhere if we aren't careful."

Owens was getting into the spirit of the matter, gesturing and hopping. Again, Ben was reminded of a lunatic swordsman looking for someone to stab. "We have to look out for ourselves. We need to talk to other miners and find out what they are being paid. We need to band together and better ourselves."

He stopped gesturing, walking normally, and talking in a calm determined voice.

"That is why I intend to look up the local Western Federation of Miners man and I will talk to him and join and I want you guys to join as well and we can make a difference if we are strong and I think..."

"Yah, Owens, do you really think?" Juni's question and tone got his attention. The big man went on.

"Stop a minute and consider: the mine owners and bankers are having the same kinds of conversations. They are not stupid, and they will look after their interests too. They'll band together, probably already have done so. And they are doing their best to keep us miners separate, alone, and easy to replace."

Ben grimaced. "Do you want to get yourself fired? Worse, blacklisted so you can't even get a job shoveling horse manure at the stable? If so, just keep talking about the WFM and how people should join it. Please, at least, shut up and think before you talk to all and sundry about wages and other mines and unions and the WFM."

That ended the talk. Just as well since they were at the outskirts of town. They stopped, shook hands, and each went his own way.

"Keep in touch, now. Good luck and we will see you soon."

III

THE THREE TRAVELERS WERE A FEW OF MANY WHO DESCENDED on the Cripple Creek gold district. On reaching town they separated, each following his preferences. Juni felt comfortable over the hill in Victor with big rich mines and lots of ore in circulation. Owens gravitated to the fringes, the town of Altman, a union sympathizing area where the Sheriff took a Deputy or two along when he visited. Ben fit in to Cripple Creek proper, where trading in mines and property preoccupied the citizens.

Like they hoped, all who were willing to work got hired. Each took a job as a miner, and each intended it as a base for other activities. In a mine the new guys always started with gofer and mucking tasks. Before long there was more to be done: shoring, drilling, equipment maintenance, and so forth. At least for Ben, life, for now, was back on a predictable and manageable arc.

Ben fell into the typical workman's routine: up early, work, home late. He still had some savings and was adding to the pot weekly. In fact he was starting to think about buying a place. Study and observing how others prospected took his spare time. He read the papers avidly, marking the claims reported on a sketch map. There seemed no patterns or obvious trends, but he kept at it. And he worked, six days a week.

The district was dotted with mines. Up close it seemed like there were more even than what he first saw up on the ridge. Many were deep, hundreds and even over a thousand feet. At depth the air was stifling and hot, even before the body heat of men and friction heat of machinery added to the stuffiness.

Ben's mine was deep, not the deepest, but plenty dark, hot, and confining. He loved the crew and the work and the satisfaction of producing ore and wealth. On the other side, he didn't care for being shut in, away from sunlight and breeze. Every day shared a small but real pleasure, to emerge after the shift and feel the fresh. Even if he had the late shift and came out in the dark, it was still breezy and unrestricted. The mixture of horse manure and coal smoke smelled extra good then.

Access to the working faces was by elevator car. Each mine had one. Typically there was a tower inside a shaft building over the hole. This housed the car and machinery, and often an office and other rooms. The car was held by cables which hauled the car up and down. Some mines used their own steam engine; some used electricity. The car ferried men and equipment to and from the business end of the mine. It was an enclosed steel platform with side bars not unlike a jail. There was a door, a gate. Cable cars ran all the time, helping the vertical commute. A car could hold up to twelve or so, its size limited by the dimensions of the shaft.

Ben's first ride made him very nervous, a common reaction. There were stories of new guys, green workers, sobbing or screaming. Of course that was a death knell to their jobs— if they couldn't take a small car's ride down, how would they handle eight or ten hours working under millions of tons of rock, in small dark passageways?

No such thing had ever happened to or around him. Before long he recognized the mine elevator car as safe and dependable. Most days he thought nothing of the daily drop (so to speak) and rise of several hundred feet.

For some reason, today's ride up after shift felt unusually crowded. He looked around and realized they had one more body than there had been on the way in. No one talked, quiet after a day's work. It felt as close as the crowds watching a parade in Denver, but certainly wasn't as festive. Riding in a metal cage held only by cables was part of the job that Ben was usually calm about. Occasionally, rarely, it got to him, and he eagerly awaited the bump of the car reaching the top.

Ben tried not to think about being several hundred feet above the floor of the mine, and as far below the terminal with its daylight and fresh air. He was hanging in a vertical rock tunnel, watching the walls drop away, and the trip seemed to take forever. He kept his mind blank.

The welcome clunk of the car at the landing woke the riders from their daydreams. A friend spoke.

"Hey Ben, want to hit the club? We're thirsty, going to have a beer or two. See the gals, who knows from there..."

Ben knew that many, including his old buddy Owens, did this as often as they could afford. For most miners, that meant several nights a week. Expenses had stacked up while Ben had toured Colorado looking for work. He finally had them paid and now was building up his savings pile. He loved the security of having money to fall back on. It opened options and allowed him to explore possibilities. He loved the job but couldn't see himself swinging a hammer or mucking a car forever. He wanted to be doing something better in five or ten years.

He liked women and partying as much as the next guy, but going out to get drunk and broke wasn't for him.

"Nah, not tonight, thanks."

"What, you gonna stay home and read again? You really ought to get out some. Visit the clubs, see the girls, drink some beer. Come on with us."

"Oh I get out, and I see people. I just find ways to do it that leaves money in my pocket."

Actually he had plans. No need to share that, they'd just keep picking at and cajoling him.

"You guys go ahead. Have a beer for me. And say hi to the gals. See you tomorrow morning."

The man put on a big smile. "Alright, Ben. If you insist, we'll hoist one for you!" He looked at the others. "Ready? Let's go!"

Ben watched them go, shaking his head good naturedly. He walked toward and past his place. In a block there was a mercantile building, a general store. It was owned by the union, the Western Federation of Miners. He thought back to the discussion he and his friends had about union stores, up on the hill before coming to town. The store part was still open with customers shopping, but he didn't go in. He went around to the back where there was an outdoor entrance to the upper rooms. There were stairs and a solid looking bannister. The stairs creaked, and he figured that was by design. Impossible to sneak up and eavesdrop on any meeting here! The door swung open as he stepped on to the landing.

"McNall, good to see you. We're about ready to start. Take a seat."

It was a big step. Of course he'd seen how owners used men and they had very little say over their pay and work. Being of farm stock he was used to doing things himself, not in a group. But he felt that miners really needed some way to influence their jobs. So he decided to look into the union, the WFM.

Ben didn't know the speaker and was surprised to be called by name.

"How do you know me?"

He stuck his hand out. "Frank Shuler, nice to meet you. I work here at this shop. The thing is, Ben, we keep an eye on up and comers. We want guys like you. You are doing well on the job and we hear you are aware of the owners' shenanigans and we welcome you."

Ben shrugged then stepped on in. There were fifteen or eighteen men and a few women sitting in chairs arranged around. On stands were a US, Colorado, and a Union flag along the walls. People seemed to be talking softly, proba-bly gossiping about claims filed and who was seeing who. At first glance no one really caught his attention. Any of the men could have been on his crew. The women pretty much looked like miners' wives or sisters, hands and eyes worn by work and worry.

One stood out. Her red hair, up in a prim but somehow alluring bun and her intelligent brown eyes caught him. She glanced at him, coolly held his eye for a moment then looked away.

As this went on, Shuler went to the middle of the cir-cle and spoke to the other, a big tall man. Then the greeter addressed the audience.

"Good evening. For those of you who don't know me, I'm Frank Shuler, president of the local affiliate of the Western

Federation of Miners, the WFM. I think pretty much everyone is here. We need to get started in any case. Thanks for
coming to this meetup of the WFM."

His unyielding gaze swept the room, and there was steel
in his voice as he loudly proclaimed, "We stand for the working miner!"

The people stood. Ben was thinking about the redhead
and noticed the movement half a beat late. After everyone
was up, Shuler led them in the opening. Three times they
times repeated, "We stand for the working miner!"

The man gestured for them to sit and they did. Some of
the men started a chant of "All for the worker, none for the
owner!" Most did not join in. They quietly looked to Shuler
to start the meeting.

He smiled but gestured to discourage the chant. "Again,
thanks for coming. We know it is a chore to come out after a
full day. This will be a good informative gathering and we'll
get right to it. Tonight we have with us the Executive Secretary of the Western Federation of Miners, Bill Haywood.
He came from the head office in Denver to meet with us. His
topic is how we can improve the lives of our members, and all
working men and women. Give a hand to Big Bill Haywood!"

So that is Big Bill Haywood, thought Ben. The nickname
seemed apt—the man was not svelte, rather beefy. He had
quick, alert eyes which surveyed the room. Ben felt he had
been noted, classified and rated in the flash of Haywood's eye.
The man had an attitude, a presence, and no wonder he had
a powerful position in the Federation. All these thoughts flew
through Ben's head as he courteously clapped for the speaker.

Haywood held his hands up to stop the applause.

"I don't deserve applause. I am just another guy trying to
do his job. If anyone has earned recognition, it is you working

miners. Thank you for the way you help create the very wealth of the earth." With that he clapped his hands, turning slowly and making eye contact with every person in the audience, man and woman alike. Smiling, the audience joined in.

"You deserve this, thank you, the government thanks you, the people thank you, and in their hearts the owners and bankers and politicians thank you!" Smiling, he led the applause for another minute or so then dropped his hands.

"Now to business. Tonight, I want to talk to you about something important which needs to be created. I'm talking about...our future. The future of the union, the Western Federation of Miners, dear old WFM. But I'm talking bigger than that. I'm talking the future of the camp here in Cripple Creek. Like he said," here he nodded at Shuler. "Like my friend your president Frank said, I'm the Executive Secretary of the Union. What does the Executive Secretary do? By the way, don't get the wrong idea just because of the word 'executive'!"

Pause, smile, a few chuckles in the crowd.

"I'm not an executive. Not a fatcat rich man like Penrose or Moffat or Carlton. I don't sit and count stacks of dollars. I don't travel around and pretend to be important while I expect other men to work. To work and make those stacks of dollars I love to count. Not me."

Ben knew these names were of mine and hauling company owners here in Cripple. They were powerful and wealthy men who had the Governor's ear. It was men like them who paid his and others' wages.

"No, I am NOT a capitalist. Not a leech. I do not believe in capitalism. The answer is not rich men using up workers like ants. That way of doing things, those practices, will end up in the dustbin of history. The men I mentioned will go down as self serving users. What is the way to do things, you ask? The answer is the WFM! The answer is strong Union action! Men—and women—binding together, relying on one another, to stand firm against being used."

He again glanced around the room, reading the crowd.

"In a few words, action to protect us workers requires quick decisions and agility."

Haywood stopped, paced, gestured, building what he intended to be a dramatic pause.

"The way we at WFM have worked has become out-moded. Democratic methods need to be adjusted! Up till now we have had member votes on all decisions and actions. That is, members have had a say on each and every issue, each and every decision that the local takes. Now, it is good that members know and care about their organization. Important that they do. But there's another side of the coin. Consider: If members pass on every item, it means nothing happens fast. Every question, every suggestion, every action is and has to be discussed, re discussed, and talked to death before it goes to a vote. Sometimes it has to be voted on several times."

He closed his eyes and made a snoring sound.

"Enough! That way is fine and dandy if we are up against a bunch of cattle. But we are not up against a bunch of cud chewing, stupid animals. We're dealing with a smart team of ruthless killers—wolves. The sad truth is, we are the cattle. And they are the wolves. Another harsh truth is, the pack is circling. I can almost hear them yipping and howling, even as I stand here."

The crowd was rapt, imagining the sound of wolves on the hunt. He looked around the circle, judging his words.

"I am here to propose that we not do that, review and vote on every issue before the local unit. We need to prepare for the struggle. I propose that we do not burden the members with every little decision and issue that comes along. That we change how we respond. We need to allow the WFM, through its officers, to act on behalf of members."

Gauging the reactions, he paused.

"Now, I am not trying to move all decisions to the headquarters in Denver. Good God, no. I propose quite the opposite. We won't have the Board and Officers in Denver making decisions about issues here in Cripple Creek, or in Aspen or Telluride or other camps. What I propose is that local officers—the guys like my friend Frank and other local officers be authorized and empowered to act."

He nodded at the president, who looked out and smiled.

"We want the men on the front lines to make the decisions. We want to empower local officers, local elected men. We want them to be able to make local decisions and to act. We need them! We need them to act, in order to bring Moffat and the other mine owners and their banker friends, that pack of running dogs, to heel."

The circled folks had been listening closely. A few had shifted in their seats. One of the women, the red haired one, spoke up.

"What about the rank and file members? After all, it is their dues which go to pay your salary. Why should they give anything up? If you get your way, what say will they have? They are the ones on the front lines. Shouldn't they have control of what happens?"

"And they will." He smiled as he said this, but his eyes were cool, even cold.

"Good question. The members will still have final say on Union actions. Any member will be able co call any question and demand a vote if there is concern. But this will stream-line things. This way, decisions and issues can be dealt with more quickly. If the Owners pull a fast one, chopping wages or extending hours or the like, we can respond effectively. We can act without gathering members and discussing mak-ing motions and seconds and finally voting. We won't need to call together the membership with all the delay that entails."

"I'm not sure I like that, cutting the members out." This from a man across the room from the woman. "I think we—you—need to discuss an important change like you propose. In public, at every local, often and thoroughly. This change you propose is a big thing and it needs to be talked over and understood. By all the members. And you need to get their consent. Specific consent by private vote."

Haywood scowled. "Well, we can talk or we can act. Can't act to take on the owners if we're always gabbing. Some meetings remind me of a bunch of old women gabbing about grocery prices or something."

Another man stood. "I am a senior WFM member, been active probably longer than anyone here. Longer than you. See, I remember when you came on board, Mr. Haywood. So I have seen things come and go. This is a big change, a big change you want. Another thing, it is the members' union, not yours. I have to say that miners discussing their lives and the future of the union is not the same as a bunch of old women gabbing. Sir."

With this last he cast a cool glare at Haywood, then went on.

"I agree, we need to take this to the members for their discussion. In due course we will let you know if we agree to this shift in power."

He looked around at the other members, turned and calmly walked out.

Who that man was, Ben didn't know. He watched Shuler take in the room, looking around to gauge the others' reactions to the old timer, then at Haywood. The big man smiled, at least his lips did, and he firmly spoke.

"Of course the members need to know about and agree to this. Take it to them, and I encourage you to talk to your coworkers and fellow unionists. They'll see the advantage. I'll be back next month and we can talk some more."

Haywood looked at Shuler who nodded and spoke.

"It is getting late and we all know the next shift comes around soon! Thanks for coming and we'll see you again in a few weeks. Meeting is closed." He banged a gavel.

The members stood, talking among themselves, slowly cleared the room.

Shuler came over and took Ben by the elbow. "Come, Haywood wants to meet you."

Shaking the big man's hand wasn't on Ben's list, at least not the first thing. He looked around, hoping to meet or catch the eye of the brown eyed redhead. She was just going out the door. Nothing to do for it since he was being hustled to the front.

Haywood shook his hand. "McNall. Good to meet you finally, have heard about you and your work in Aspen."

"Oh? I was just another miner until the bottom fell out over there."

"You worked well with other miners and even had connections with a few of the owners. That is a good background for you to help us here. So, what do you think of the idea? Of local officials acting fast?"

"It sounds good on paper, Mr. Haywood."

"Call me Bill."

"Alright, Bill, thanks. Like I said it sounds good on paper, but it goes against the founding principles of the union. They have been responsive, that is members have, when there is a problem or challenge thrown out by the owners. Why do you need to fix it?"

"Because the owners are fixing to clamp down, try to cut wages. Or I hear they are thinking on it and we want to be ready."

"They'll never get away with it. That is a perfect example—if they try the miners will down tools and block the entrances. Within hours. No need for discussion or formal vote, no need for the local officers to try to give orders."

"Well, Ben, times are changing. Things move fast and we need to be able to respond."

"I'll give you that, Bill. It is important to keep up with the times. You know this of course: We miners are independent as a hog on ice. We think for ourselves. Give it a month or two for folks to think it over and see how it would work for them. I expect the members will come around. At least part way.

He grinned. "Us men who grapple with mother earth every day will be slow to give up any say so on anything. Don't push too hard."

"Good point, McNall. That orneriness is an advantage and a strength for miners. But sometimes you can overthink things. Sometimes you just have to go with a change, no?"

Ben shrugged. Damn, he was thinking, I wish I was talking with that woman instead of this big union ape.

Haywood of course didn't know that. He was still talking.

"Have you eaten? Join us. I'd like to get your views on the district. On the mines and owners. On how we can make this a better place to work and live."

Off they went and took seats at one of the taverns. It was doing great business. With so many mines there were always hungry and thirsty miners coming or going. Ben wondered if work conditions and life were so bad, why were this and other clubs going great guns all day and most of the night? He stuffed that thought and paid attention to his hosts.

The union men compared notes on goings on here and in other mining camps. Ben listened more than he spoke. Haywood and Shuler talked on.

Of course more beer meant more and freer talk. They speculated on owner's concerns and actions, union strategy, tactics, the advantages of union owned stores, wages and hours, conditions in other mining camps, how the membership viewed them, and so on. It was all business. Ben had expected at least some light conversation, but no, this was a working dinner. Haywood was a serious man on a self imposed mission, and there wasn't time for personal talk.

"And I have to remind our people sometimes," here he glanced at Shuler. "Not you, Frank, but some of our line members. I hate to say it, but this isn't a bridge game. We aren't in a game that is all polite and structured according to some rule book. This is a labor union, fighting for rights and the lives of our members, and to grow. For us to do this, from time to time people have to be 'encouraged' in an assertive and unforgettable way." He paused, looking around at them. "I didn't say this, and will deny if you repeat it: Sometimes, heads need to be broken."

Big Bill said this enthusiastically. Ben got the impression he not only didn't hate to say it, but that he enjoyed getting involved in such 'encouragement.' Come to think of it, the man's knuckles were scarred and the nose looked like it had been broken a time or two. Ben half listened as he considered this.

"Yes, there are times when only a shock can bring people around to our side. And there is no middle ground, rather two sides. Either you are for labor or you are a capitalist. What was it Thomas Jefferson said, 'from time to time the tree of liberty must be nourished with patriots' blood.' Well, sometimes the tree of labor and the Union must be fed with strike breakers' blood. Sometimes we need to pull off the scabs."

Haywood's leer at this awful play on words made Shuler laugh, but it sent shivers down Ben's spine. "Now, I don't mean random or brutish beatings, although like I say there is a use for fists now and then. Occasionally the use of light force will send us some of the weak, the fearful, although we don't really want those types. Leave them to the owners."

He winked at this, and finished the thought. "What I mean is sometimes there is need for a bold action, intended to make a statement. Something that even the owners can't mistake."

Ben mildly observed, "Can't most of our objectives be gained at the negotiating table with both sides working in good faith? I mean, everyone wants things to be peaceful and fair, right?"

Haywood eyed him coldly. "Maybe I misread you, McNall. Surely you know those capitalist bloodsucking owners have no faith, no honor. Any action or agreement they take is purely for their gain, workers be damned. Never forget that, Ben."

"Well, Bill, you have more experience with them than do I. I have to go with that." Ben yawned. "It is late, and like Frank said at the meeting, the next shift starts in a few hours. I had better go. It was good to meet you, Bill, and hear about the WFM's workings. Thank you for that, and thanks for dinner."

They stood and hands were shaken all around. Haywood looked Ben in the eye. "I'll be back in about a month. We'll get together and talk. In the meantime, speak to your mates about my ideas—they'll see the light!"

THE NEXT MORNING, BEN RODE THE CAGE DOWN TO THE WORK-ing level. It didn't feel as crowded this morning as it did on yesterday's ride. He eyed his 'mates' as Haywood called them. Was their lot bad, or were they fortunate to have a job? Were they fairly paid? Yes and no, he decided. He was happy to have a good job—safe mine, good crew, decent pay. That said, conditions could certainly be better. He didn't agree with Haywood that all owners were selfish and evil. But, as long as they held the whip hand miners would have to put up with the job as the owners wanted it done.

"Hey guys, how was the sporting club?"

Through bloodshot eyes and with dry mouth, one of them tried a grin. It looked like a river crocodile at the end of the African dry season. It momentarily made Ben uncomfortable and thirsty. What a grimace!

He croaked, stopped, cleared his throat and tried again.

"Oh it was a good time and the girls asked after you. Why we almost drank the bar dry, didn't we, fellas? I tell you, there were a lot of miners in there, most having fun like we did but some were complaining about their pay and hours and how they had to go in early and stay late. I tell you I got so damned tired of their carping when they should have been having fun, I almost took a swing at one of them. And get this, the Union is thinking of taking our vote away, can you imagine that? I tell you they had better not try. Next thing you know they'll tell us where to shop and which newspaper to read. Not to mention, which girl you can choose as a friend and who we can go have a beer with. I don't like it I tell you."

"Oh?" Ben was fascinated that Haywood's idea had already hit the street, or the shaft so to speak. He probed. "I'm not sure what you mean. Take the vote away? How could they do that? And why would they want to?"

The cage stopped and the door opened. A foreman stood out on the landing.

"Alright you men, come on. There's ore awaitin' for your tender love! You have tools to set up, muck to move, holes to drill, shots to set. Forget the gals from last night and get with it! Sure, it is a nice August day up top but we can't mine sunshine. Don't matter, like I told you, we have ore to handle."

A friend muttered, low and quiet like, "Who the hell is this 'we' he talks about—I'm not sure he knows which one is the business end of a shovel."

The boss turned and strode away, not even looking back to see if they followed. The man was busy allotting and pointing out tasks, dropping one or two men at each.

As he followed, Ben thought of the three dollars of pay he would rack up for eight hours of work. The hours of the work day were really ten or more since he had to set up at the start of shift and leave things just so after shift. That was expected, and the time so spent wasn't considered part of the eight hour day. It had always been that way, far back as he and everyone else remembered.

Soon he was hard at it, working with practiced motions but always alert, always aware of the motion and sounds around him. At work, a miner couldn't ponder pay scale or Union politics or fun at the sporting clubs or anything else. Not while on the job. He had to concentrate if he wanted to stay alive and uninjured.

After shift, going up in the cage, one of his mates looked over at him. It was the guy who asked about the foreman's 'we.' Smiling wearily, he resumed the conversation of the ride down. "I'm tired of working but not being paid. If the Union can get me more money or less hours, or pay for the hours I actually work, all to the good. Now that I think on it, I would willingly let the local officials have my voice to do that."

Ben, not wanting to color the guy's opinion, paused before replying. "I see your point. Personally I don't like to give up my right to have a say. But I know what you mean about work without pay. That just isn't right and however they do it I hope the union changes it."

The cage bumped to a stop and they filed out on the surface. They were inside the lift building which was more or less a shed with a few rooms, open sided on two direcctions. Outside they could see that the sunlight was a washed out, weird orange. A faint smoky scent was all around. Ben wondered what happened to the nice summer day.

"Fire in Victor." The operator glanced out at the haze then back at them. The man was known to be economical with his words and lived up to his reputation this afternoon. He added nothing, just went back to work, waiting to see if anyone needed to ride the cage back down. He said nothing about when it started or how big or which mines were threatened or was the town itself in trouble, or anything else.

Outside the lift building several men gathered, looking out at a big vent of dirty brown smoke. It rose hundreds of feet, sitting over an ugly glow. The base of it shifted directions, dancing around as if being pushed here and there by an invisible hand or something. It veered several different directions even as they watched. Ashes fluttered, a flurry of white in the summer afternoon.

One of the men reminded Ben of Owens. He was nervous and twitchy, same same. The man shifted his stance, unable to be still. He burst out with a staccato soliloquy on fires in gold camps.

"Not good. Not good at all. Hope no one is caught in that inferno. But you know, Victor was due because Cripple had a big fire, two within several days, several years back. April of '96 it was."

They looked in wonder at the dark column. There was a distant dull roar, with occasional pops like .22 ammo cooking off. Probably it was.

"The first one back in Cripple ran its course and got put down, and everyone was relieved. Some even started picking through the ruins and began to rebuild. Then four days later another fire got going. That one nearly wiped the town out."

He stood still for a moment, looking at Victor's funeral pyre.

"All those wood buildings with lanterns and cookstoves and cigarettes and so forth, it was bound to happen. It always does in every mining town I have heard of. Cripple's fire was awful, lot of folks lost everything and a few died but it cleared out some of them shacks and so forth and the town rebuild and is the better for it. Hell, even the people of Colorado Springs donated food and materials and help. So did Leadville and other mining camps too."

The miner paused, done with the story of Cripple's fires. He turned again towards Victor. "I wonder how this one got started. Bet it was a kid playing with matches or something. I better go see if they need help."

He rushed off, his awkward but efficient movements reminding Ben of a lobster chasing a starfish. The guy was intent, excited, focused, and totally absorbed in the task at hand.

Ben went home. He figured plenty of people were offering help and he didn't need to crowd the scene any more. That evening he read in the paper that hundreds of Victor's buildings and at least eight square blocks were burnt. Some people were still missing. How it got started was being looked into. The reporter said that the flames seemed to have started at one of the sporting clubs. Some kind of accident was suspected.

Blearily he tossed down the paper, thinking of his short night. Listening to Big Bill Haywood wasn't as entertaining for him as his mates' foray to the clubs for drinks and women. Even so he figured he was just as tired as them after a day's work. He yawned and turned in.

IV

In the early days of the Colorado gold rush, the stuff was relatively easy to get. A man coming early to a strike could pull nuggets out of a brook's riffles. Gold flakes could be panned out of the creek. Gravel could be washed down a sluice and gold found in its built in small baffles laying shinily, heavy, for the taking. As in all gold rushes, the bonanza soon ran its course. The loose, unattached gold was found. No longer was it simply there for the taking.

The easy gold may have been gone but the desire, the need to have it did not subside. If anything, more effort was put forth. The treasure wasn't to be had by simple efforts but ways were found to get at it. Gold and other minerals were interspersed throughout and locked into rock, hard rock. Physics and chemistry were put to use. By trial and error, knowledge needed to wrest the gold from the earth's rocky interior grew. Machinery and specialized methods were invented and improved. It was the only way to squeeze wealth from the earth. Still, expensive machines needed men to run them, and to quarry the ore to feed them. Thus evolved a partnership, men to dig and machines to process. For decades it would provide immense wealth, many jobs, and much turmoil.

Thus the dance of labor and capital. It was playing out across the country in many industries. The tango involved the engineers, brakemen and railroad tycoons, assembly workers and craftsmen working for a factory magnate, roughnecks and the Rockefellers as well as miner and mine owner.

A couple dancing the tango look gliding, passionate and adoring. In the arena of business and labor, the partners' steps were sometimes smooth and smiling, usually sullen with one partner grudgingly following but grabbing the lead whenever the opportunity came. Sometimes the dance became outright hostile. Then the music stopped and the partners stepped back.

Between work, his reading and conversations with Haywood and Shuler, Ben saw this uneasy arrangement writ large in Colorado. From the early 1860s on, Colorado was a hotbed of mineral development, knowledge, and machinery. Miners' attitudes and practices were shaped accordingly. As mentioned, it didn't take long for the panning and sluice work to exhaust the surface gold. When that happened, the industry almost stalled. The ore dug from the earth remained locked up, the riches in it inaccessible. Somehow it had to be crushed and processed to get to the good stuff.

At first no one knew how. Before long men stepped in and adapted ways from other mining areas. Elsewhere, complicated industrial plants needed consistent power, water, transport and other resources. These needs and facilities called for big investments of time and money. Also it called for a large supply of trained and reliable workers to feed and operate the plants. And this army of workers needed to be managed. They needed to be organized, the work defined with tasks put in the right order, the machinery maintained and properly used, safety ensured, the men paid.

Colorado's mining industry was in the doldrums, almost dying on the vine. Long story short, a man named Hill figured out how to economically separate gold, silver, copper, lead and other metals, all locked and interspersed in the rocks, the ore. Hill's methods and activities unlocked and created a dynamic and important industry. Untold wealth was created and that in turn provided jobs for the burgeoning population of the western US. A byproduct of the wealth and gold from Hill's methods was big piles of tailings, dross, at every mine location.

This was in the late 1860's into the 70's, when the industry was slowing dramatically. Hill had a background in geology and engineering. He visited other smelters including state of the art facilities in the United Kingdom. He combined what he learned from others and applied it with his knowledge of Colorado's unique ores. Actually he started with the ores from Central City and Blackhawk. His vision and the processes he pioneered in Colorado enabled mining to become the driver of the state's economy for many decades.

Which came back to the need for men who were able and willing to go to work in mines. The days of independent prospectors or gold panners making it big were pretty much over. Men hired to work in mines had little choice. They could work the long, very long odds, of making a strike on their own, or they could take a job. That meant accepting wages and conditions offered by mine owners. Miners and prospectors, being independent and ornery, didn't care for having little say. Many fought this.

Hill's techniques and discoveries fostered consolidation. Over time, mine owners often bought or built the mills to process their and others' ores. More and more power was concentrated in a small number of hands. An owner of a major mine and mill could control or heavily influence the entire mining camp. Sometimes this influence was benign; often it was not.

As in other US industries at the time, the workers reacted by organizing unions.

Ben knew there were whole books written on these subjects, mining and smelting and labor and division of economic power and unions. He also knew that the mine owners often had the ear of governors and legislators. The owners seldom hesitated to ask for the militia to intervene in labor disagreements. In point of fact, the mine owners practically viewed the militia as their private police force. This was common, bitter knowledge among the miners. All one had to do was read the papers.

The history and technical aspects didn't really matter. Ben and hundreds of other miners cared about real issues, day to day living: they wanted a fair wage, to be paid for all the work they did, and a to have a reasonably safe workplace.

So, here it was, turn of the century. Ben was in Cripple Creek, Colorado. He and many others were working in the mines and happy to have a job. There was not a general dislike of working conditions—most seemed content. But still...
He had read of troubles in the Coeur d'Alene mining camps in northern Idaho back in the early nineties. And he knew men who had been in Telluride, over southwest. The militia had been involved there in 1893 or so, and things got real ugly. That was back before the silver crash. Things got worse

everywhere afterwards. And then there was the strike of 1894 right here in Cripple, the battle of Bull Hill.

He thought about that occurrence most every day, walking to the mine or riding down the shaft. That whole scenario, Bull Hill, must have been something. He even went to the local papers and read up on the newspapers of the day. It was a series of events, not a one day happening.

Bull Hill was a hill near Victor, away from the town of Cripple Creek. Of course mostly miners lived there, and the settlement up there was definitely pro labor. The WFM called for a strike in '94. The demand was for three dollars of wage for an eight hour day in the mines. Not everyone saw the need to strike. In any case, the call went out and the ball got rolling. The miners up there were fortified, armed, and skittish as hell. They literally had a fortress and kept it manned and armed.

As background: The Governor, Davis Waite, bless his soul, was a liberal, a progressive. He was elected to a two year term in 1892, serving 93–95. He was not in sympathy with the mine owners and other moneyed interests. One of the things he is remembered for is corruption. Or the lack of it around him: he had tried to clean up cronyism and corruption in Denver city government.

He actually tried to get rid of, eject from office, some Denver cops and firemen; they declined to leave. Not only did they decline to leave, the occupied city hall. They literally took over the city administration building, took it and holed up there. They hoped to spark a confrontation and gain credibility, or at least to keep their jobs. Ben had seen photographs and sketches of uniformed men leaning and gesturing out of windows in Denver's city hall. That confrontation,

Waite versus city cops and firemen, never came to violence although it was apparently a near run thing. Ultimately cool heads prevailed and the sides started talking. The Governor prevailed, more or less. He was able to get rid of some of the city employees.

This Denver comic opera took place at the same time the WFM was stirring up the miners in Cripple. The guys up on Bull Hill were getting restless. They went out on strike for that three dollars/eight hour day. The mine owners took exception. They tried to keep the mines open with non union labor. They also saw the need to meet force with force—if the miners up on bull Hill wanted a fight, the owners would give them one. They went to Denver and hired some of the men Waite had gotten out of city hall. Ex-cops and firemen went on the payroll of the mine owners, as security guards. Their job was to intimidate the miners back down the shaft, back to work.

Ben mused on the scenario he had read and heard about. Must have been something, tense and rumors flying and people coming and going. It was almost like, what was that British outfit? There were two of them who wrote operettas about the Royal Navy and the upper crust and all that. Gilbert and Sullivan. That's the tag he was trying to remember, Gilbert and Sullivan. Reading about it years later, the whole thing could have been written and staged by that British duo.

There actually was a confrontation, the so called Battle of Bull Hill. It wasn't a battle in the Waterloo mold, not even really a skirmish. The 'Battle' unfolded as if per a script by those two English composers.

All the elements were there, except for no singing: The ex Denver cops were deputized by the local sheriff. They swaggered and blustered and flaunted their authority. The Militia was called in and the soldiers too were happy to throw their weight around, backed by rifles and Gatling guns. Trains were coming and going. They hauled not only the normal merchandise in and ore out, but also striking miners and law enforcers. Potshots were taken at some of the trains. Just who pulled the trigger was never established. One train full of ex Denver men reversed course when it was opened up on, and the would be deputies left town in a hurry.

On the serious side, the strikers were ready to defend their fortress on Bull Hill. Everyone knew they would shoot if need be, or if they felt threatened. That was beyond doubt. There was tension aplenty, and drama to suit a Broadway star. Comings and goings were seen all over town, any hour of day or night. Meetings were planned, orchestrated, moved, canceled and re scheduled between owners, union big shots, the militia and the Governor. In the end, the miners got their eight hours. To the relief of all, there were no big gun battles. The town and everyone in the district got back down to the business of making money.

Now, eight or so years later, things in Cripple looked similar. Same song, second verse—the WFM wanted more money and the owners wanted rid of the WFM. Neither side had a real beef but both were looking for a reason to lash out. Ben was especially aware of this on the ride down that morning. Some of the men were twitching and bitching more than usual.

He also knew that Big Bill Haywood and local Union biggie Shuler were getting together to stir up the action, get the members hostile and excited. The big man worked to get his local control measure through. While he was pushing for a strike he was also trying to limit what the members heard and read. Last thing he wanted was for members to realize the problems they faced were in fact small. If they knew that they would likely tap the brakes and Big Bill would not get his strike. As it stood, WFM members were mostly reading and believing what he said. They had little say on the course of events.

THE NEXT DAY, AFTER SHIFT, BEN RAN INTO HIS OLD FRIEND Juni. The affable grin seemed bigger than the man, hard as that was to imagine.

"Hello, Ben. How goes the battle? I haven't talked with you in a while."

"No. Been busy. Work, union business, life, you know how that goes. Juni, I have to say, you seem happy. Things are going well for you, it looks like?"

Juni looked around to be sure no one was eavesdropping. His grin grew.

"Yah. Can't complain. In fact yes, things are going pretty good. You remember what I told you about high grading?" Another dramatic look around and the stage whisper he used for the last words made Ben laugh. Juni was not amused.

"It isn't funny, Ben. I take risks, sure, but it makes me money, good money. Better money than I ever pulled in trying to be a silver miner." He stuck a hand in pocket and partway pulled out a wad of bills, riffling them and raising his eyebrows before sticking them away.

"Plus I have met up with some folks who really know the ropes. I mean, there are many ways to run this business, some would never have occurred to me. Have been fortunate to work my way in. Literally! I'm learning of old abandoned access shafts. My God, there are hundreds of them here in Cripple and Victor! You can get into most any mine you want. And there are so many places to stash a load to come back for when no one is looking. And I have gotten to know people I can move the ore through."

He stopped, glanced warily around and continued. "I even met a man with experience in Coeur d'Alene. Up in Idaho, another big booming mining camp. He was involved in the strikes and deportations up there in '93. Got cheated out of a bonanza when the governor ordered him and others out of the state, and he is mad about that. Anyway, he knows the lay of things here. The man goes by Harry Orchard but I believe that is an alias. Whatever, the guy is a master. I learn lots every time he takes me out! New ways to get into and out of mines unseen, sources of good ore, and other things..."

"Juni, I'm not sure you should be telling me this. Sure, we all know miners and even some foremen take ore on the sly and sell it. It is not a secret. But do you think it wise to name names and get specific, at least out here in public? I don't want to know. Just don't tell me any more. In fact, I heard the mine owners have organized to stop it. Or at least control it."

"Nah, they don't care. As long as it isn't blatant. There is plenty of ore to go around. I'm comfortable telling you about it. I trust you, Ben. These are good folks. Especially Harry—I like him."

"Well, best of luck with all that, Juni. Like I say, this is too much information. More than I want to know." Shrugging, Ben continued. "But it is great to see you happy and feeling good, I have to say. Well, my man, I have to run. Buy you a beer next time?"

"Sure, Ben. See you around."

CRIPPLE CREEK AT THE TURN OF THE CENTURY WAS MORE than a mining camp; it was a city. There was no central accounting or management, no listing of citizens, no registry of population. Cripple Creekers paid no more attention to registering or checking in than did dandelion seeds in the wind. Still, the district's population was estimated at some fifty thousand. People arrived, sure of making a fortune. Of the arrivals, some left after a week or a month, out of money, energy and hope. Most got along alright, making a living. A small group, very few, did in fact find wealth.

Some came with no more than a good attitude and work ethic. Also, there were the inevitable users. Some men and women came to work the clubs, taverns and poker tables, not the mines. They chose to harvest wealth from the miners not rocks and ore from the earth. Merchants came to provide legitimate and needed goods. And there were some who came prepared: Of the many arrivals, a small number had financial support and mining knowledge, theoretical or actual, to back up their optimism and willingness to work. This day one of the latter arrived in town.

THE MIDLAND TERMINAL TRAIN CAME FROM THE NORTH, ITS line connecting with the Midland's main line west out of Colorado Springs. As it came over the hill, Dillon Bosini looked out the window. He saw a big bowl high in Colorado, next to Pikes Peak. And it looked as alive and busy as the ant colonies he used to kick over as a kid. It literally crawled with activity.

Hundreds of mines had men, ore carts and wagons coming and going. Smokestacks spewed hot air and steam and cinders. Houses, stores, sporting clubs, churches, schools, electric power poles, train tracks and stations, and people, many people of all backgrounds and appearances, all contributed to the general delightful and voracious ferment.

Lon—no one but his mother called him Dillon and at a young age he insisted he be called Lon—was excited. The moment was as he had thought it would be. As the train neared the station, he could almost feel the gold and enjoy the coming wealth. The place reeked of money to be made and it felt dynamic and hurried and exciting, growing and important...

Briefcase in hand, Lon waited his turn to leave the car. Soon he descended onto the bustling platform. To the casual onlooker, his blue eyes with dark hair and complexion were somehow not noticeable. The combination was one that folks didn't really catch at first. After a while, later, people realized why the unusual combination made his looks unusual and striking. The newcomer's nice coat and well knotted tie told that he was not just another rock knocker hoping to get on the payroll at some mine.

A WOMAN RUSHED UP. "LON! SO GOOD TO SEE YOU!" ALMOST A year had passed since his sister left to teach in Colorado. Her red hair and brown eyes were the complement to his blue eyes and dark skin. Italian and Irish parents played true. She was unchanged, but somehow carried herself as if she was older and wiser. It was good to see her.

"Abigail!" He almost hugged her, but restrained himself from such a public display of emotion and affection. Putting hands on her shoulders, keeping his arms pretty much straight, was as far as he went. He looked her in the eye, smiled.

"How was the trip? Isn't that ride from Colorado Springs beautiful?"

"It was an easy ride, thanks, and yes, riding halfway around Pikes Peak really is something. I have to say, I'm glad to be here." He stopped and looked around again.

"So this is Cripple Creek."

She giggled, remembering her reaction to the hubbub and smoke and wealth being showed off. He came back to the present.

"Now, you! You look fabulous, Abby, and happy. Teaching suits you, I'd say. It must be fulfilling and fun. Watching your students learn the ABCs, addition and subtraction has to be good."

"Yes. Watching and helping, that is the real satisfaction. Showing them the door and helping them open it the first time. After that, most of them run to and through it, looking for another challenge. Their excitement is worth all the work. There's more, and I can't wait to tell you about it."

They walked to the baggage car and he made arrangements for delivery of his trunk.

She asked, "Mother and Father? How are they? Do they like their new home? Oh, it is so good to see my big brother! Tell me about the trip out to Colorado and what you have been doing for the past year. I'm so sorry about...the wedding being called off."

Lon couldn't stop looking around. This place made him optimistic and happy, and for the first time in over a year he didn't care that his wife to be had backed out after a date had been picked. At that he dropped his work. He had a long held dream to be a miner, so he came to Colorado, studying and learning to do just that. His never to be father in law had refused the return of the dowry. After a second try, Lon won it back and used it to finance his Colorado trip and training.

The frenzy and excitement of the world's greatest gold camp tore at him, bored into his being, made him jumpy and energetic and ready to go. In the back of his mind he again saw the view as the train came into town. He thought of the hundreds of mines, thousands of buildings and people everywhere looking for gold. Not to mention the many looking to separate gold from the prospector, miner or mine owner. The idea flashed through his head that the real way to riches in a gold camp was to sell groceries, supplies, or companionship to the miners. Maybe he should try to do that? No, he wanted to mine. There were many ways to amass fortunes; he firmly chose that one.

He repeated himself. "The ride was alright. It went well, only a few hours from Colorado Springs. It sure is good to be here, to see you, Abigail. How about we deposit my bags at the room you got for me. Thank you for that. I imagine rooms are at a premium here. Then let's go eat."

Waiting for the meal to arrive, they talked.

"I do like teaching and I like doing it here. The people here are from all corners of the globe, all walks of life. Not every child attends school, but those who do really want to learn. It is so joyful to me that my children love learning, coming to class, and reading books. Very few are there just because Momma says so, and they don't last anyway. I have a friend teaching in Chicago and she has some students who don't care to learn, but act up and get in the way of those who do. Can you imagine? If a young man here doesn't want school, fine and dandy. Let him do something else. They want to be out prospecting or mucking in the mines. More power to them."

"I'm glad it is working out for you, Abby."

"Yes. And this camp. I love it. It is a fascinating town— I mean, the owners run the place. The mine owners I mean. The owners of the big mines, not every Tom Dick and Harry who has a claim that may pay a little. The big mine owners have the organization, the money and the power. I am glad there are a lot of them, fifteen, twenty, thirty?"

She looked inquiringly at him; he shrugged.

"I'm not sure, but there are quite a few. If there were just one or two owners in town, not good. They could, would get together and agree to pay only low wages and allow hard conditions. With many owners, that doesn't work. Too many voices so there is no collusion, no back room deals on what to pay a man. And there is competition for good workers so the pay and so forth is not bad. Even so, the miners and the workers are always trying to claw power and control back from the other."

"Oh? Gee, little sister, you talk like an expert on the subject."

"Oh, I guess I am just interested. And don't call me little, buster." She didn't look him in the eye.

Lon almost frowned but willed a smile. "Sorry."

"I mean it. I have lived here on my own for a year now and don't need you to condescend."

"Really, I apologize. But seriously, Abby. I know these are hot topics here in Cripple and it is easy to keep up on them. Are you just reading about all this in the papers? Or is it more? Are you involved somehow? You aren't out talking to miners and Union people are you? You can't be believing them. You aren't buying into that union. Are you?"

"Now Lon, I'm a grown woman. I have my own place and a salary. Well, I share an apartment with another teacher, but still, I'm not a schoolgirl. I can talk to whoever I like. Even if my big brother disapproves."

"So you are talking to people in the labor movement."

"Well, the kids, especially the older ones, talk. Usually between themselves and I overhear. Some twelve year olds really understand. Plus I talk with parents, mostly about their children but things are said. Some are pro union, some not. That got me interested, and then one of the mothers told me about a union meeting. I went out of curiosity."

He paused, thinking. "If you believe in them there must be something there. Please fill me in, I'd like to hear what they have to say."

Dinner arrived. "Roast chicken?" The waiter looked between them; Abby nodded. "Here." He unceremoniously put her plate down, slopping a bit of gravy. "Here's the steak, medium well for you." The steak looked rare to Lon but he was hungry. "Anything else I can do? No? Enjoy your meal." Off he stomped.

Abby smiled. "That is pure Cripple Creek. There is a man who, clearly, would rather be out looking at rocks. I'll bet he doesn't stay here an hour after his next paycheck! I bet he goes out every spare minute, after work and days off, to look for color. I bet he loves knocking rocks. A man has to eat, but he doesn't have to work a hated job for too long, right?"

Lon already had a mouthful of steak, and merely nodded.

She giggled. "Gold fever makes people do strange things."

He choked the less than tender mouthful down. "Speaking of that. I swear, Abby, even as I stepped down off the train, I felt it."

"What?"

"The fever. This town runs a fever like you say, gold fever. This place feels, smells, has a dynamic aura of excitement. Everyone is in a hurry. The possibility to get big time rich is here, it really is. It may be just out of reach but it is there for anyone. I know this is America and all, but this place is special."

She pulled a drumstick off and took a bite while he continued.

"That waiter could be a millionaire next week. And that makes this place, this gold camp of several towns in a big old volcanic bowl extraordinary. This entire gold camp is afflicted, every man and woman here."

She nodded. "I told you about some of the young men in school, those who may show up once or twice, usually reluctantly. They stop coming and most of them have gone out to find their mine and fortune. Not only students. I've seen professional men, lawyers or architects or doctors with that fever. They arrive and soon throw everything over. I mean everything, their career and even sometimes family, to go prospect, or work a claim. It is truly fascinating."

"Well, I understand. I felt it right away like I say, and I still feel it when I go out on the street. I don't want to go do something foolish like that, abandon what got me here. But, the past year I have studied with geologists at the School of Mines in Golden. That is quite a place, almost as exciting in the search for knowledge as it is here for the search for gold."

His sister rolled her eyes and he grinned.

"Well, maybe that is overstating, but it is an interesting place. And it has a lot of practical knowledge floating around. Anyway, I hope that I picked up some of it and have a good idea of what to look for."

He cut another strip of steak and chewed it, thinking of things he saw as the train came down the mountain. "And I have to say, I saw traces of color several places while coming in. Maybe they are traces no one has noticed! Some have been, and some, people just haven't looked at in the right light with a trained eye. Who knows? In any case, I intend to go have a look. Maybe even file a claim."

"Oh Lon." She smiled, thrilled but apprehensive. "My brother really does have the fever. May God have mercy on his soul! Say, we haven't caught up on the family yet. How are Mother and Father doing? Are they well? And the rest of the family?"

Shortly, two men came in. Abigail was busy with family affairs and at first didn't notice them. As they walked past, she did. Lon saw her make a small twitch and sit up a little straighter. Then she brushed her red hair back. He didn't like that his sister had subtly signaled 'notice me' to these guys. This all happened in a quick moment.

"Miss Bosini, hello." One of the men stopped, smiling at her then gave Lon a not hostile but not friendly once over.

"Mister Shuler, nice to see you. Meet my brother, Dillon Bosini. He just arrived today."

Lon smiled, stood and extended a hand. The men could have have been prize fighters at the opening bell. Sometimes people meet and dislike each other on sight, and that was this. Lon's smile lost warmth but remained, like a wilted rose. Abigail completed the introductions, looking back and forth as she told one about the other.

"Mr. Shuler is President of the local unit of the Western Federation of Miners. Lon has just finished studying at the School of Mines, and had a good trip up from Colorado Springs today."

They shook and Lon sat back down. Shuler looked at Lon. "School of Mines, huh? Looking to open a mine are you?"

"Isn't that the dream of most everyone who comes here?"

"I suppose so. Most never get to open one, but many do get to work the mines."

Shuler turned to Abby. "You may want to come to the next meeting. Bill Haywood is coming to talk with us again. Will follow up on the changes he proposed last month. And give us updates on the MOA, what the mine owners are up to. Hope to see you then."

"Yes, I will be interested to see how the members view his changes. And his read of what the MOA is doing these days."

Lon followed this with interest since he had heard of Big Bill Haywood.

Shuler smiled. "Yes. Me too. It was nice to meet you, Lon." He nodded and walked to his table.

Lon followed with his eyes, making sure he remembered Shuler and also his companion, whoever he was.

"So you have met Bill Haywood. Is he as big and scary as the papers make him out to be? I understand he is a big man, literally. Can't say I'm surprised you know those guys. No wonder you know what is going on in the labor scene here! And here I thought you were just reading and talking to your mothers at school. Typical Abby, if you're doing something, do it big."

"Bill Haywood is not scary. Big he is, yes, and he is passionate. Passionate about workers and their pay and conditions. Passionate against owners getting it all. No, he's not a scary man, just a big, passionate, and dedicated one."

"Dedicated to what? Prosperity for all or the destruction of private property?"

She was defiant. "Miners are getting the shaft here, no pun intended, Lon. Owners make big money from the sweat of men working hard underground. They sit in offices and cash checks while men go underground and risk their lives to bring up ore to be refined. And those owners are not only rich but powerful. They have the Governor in their pocket. So they get to make the rules, and break the rules if they want. Meanwhile miners are paid a measly three dollars for a day's work. And they have to work before and after for no pay. While the Owners, at least some of them, make millions. Millions! It is not fair."

"I understand that, Abigail, my dear and precious sister." The mild response made her pause. He went on.

"The thing that concerns me is, neither side is all good nor all bad. That said, it is plain that sooner or later, a showdown between the WFM and the owners will come. You can see the signs everywhere. And I don't want you to find yourself stuck in the middle when bullets fly."

"Bullets? Surely it won't come to that? Will it?"

"There was shooting in Coeur d'Alene and Telluride eight or ten years back. Idaho Springs and Central City have seen unrest, blasts, bullets. Not to mention in the Haymarket strikes in Chicago, and many other places. Those aren't mines, but they sure involve owners and workers staring down a rifle at each other."

"I hope you are wrong, brother."

"Me too. But think, a few shots were taken in the Battle of Bull Hill just down the road here, not ten years ago. Not to mention there is plenty of dynamite around here and someone could use that to blow something or some people up."

"But...that wasn't today's Cripple Creek. We get along and we care about each other here."

"Ha! Ask your Bill Haywood how much he really cares about mine owners and managers and foremen. And the property they have spent their own money on to build and improve."

"Still, I can't see how violence can erupt."

"Abby, there is no reason it couldn't happen again. It has happened here and could, probably will, happen again. Wake up."

"Don't start your big brother act again, Dillon Bosini. I don't want to hear it."

"Pretend I'm a newspaper reporter, then. Someone you barely know but who knows what he is talking about."

Her look was as if he had lost his mind.

"Abby, I'm not trying to run your life. I'm just concerned, that's all. Be careful. And know that these Union men—and the owners too—will go to great lengths. These guys, both sides, play for keeps. They don't do things by discussion and majority vote and Robert's Rules of Order. Anyone, I mean anyone, who gets in their way can get hurt."

Eyes blazing, she justified herself. "Here you are, first night in town, and you're an expert on the owners and the union here. What gives you the right?"

"You're not the only person interested in this, who knows people involved in the situation!"

"Maybe not. But I can talk to whoever I want. About whatever I want."

"Hey, talk to who you want. I'm not trying to stop you. Just look deep and be aware of what they're really about, is all. You and I are family. We don't have to see eye to eye on everything, but we need to care about each other. That's all I am saying."

She was done with him for now. In a breach of custom, she motioned for the tab. Usually that was up to the man. She didn't care. Standing, she was ready to leave. "You must be tired. Let's get you to your room."

V

Miners treated the cage rides different depending on the day. The rides down to and up from the working face were sometimes quiet, sometimes chatty. A tired man might say little and some never said a word, ever; some wouldn't shut up.

It was a chance, if you wanted it, to have conversation and trade news and gossip. The trick, of course, was to distinguish between the two. Ben sifted nuggets of fact from dross of conjecture and meanness. Some men gave more of one or the other and most everyone knew which was which.

"I hear Jim Argyle staked a claim over on Raven Hill. Says he found color."

"Raven Hill? There are some big strikes over there, rich lodes. I thought that hill was covered over twice with claims."

"I thought so too but it is a big mountain. Lots of rocks over there. Maybe ol' Jim is on to something. I don't know, just know what I heard. His friend Cooker Eisner told me. Oh, say, did you hear the one about the mine owner and the mule?"

Listening, Ben wondered. He knew Cooker. He knew Cooker was out knocking rocks every day, early and late, prospecting the heck out of things. Cook trusted Ben and had confided some of his work and exploring, as much as one prospector was likely to share. And Ben knew that none of Cook's recent work was near Raven Hill.

In the cage of course he said nothing about this. Seemed likely that the story about Argyle and his Raven Hill claim was a smokescreen. A lot of folks did that, trying to protect their ideas and claims. Thus the sifting of gossip and 'news' for the truth.

IN ANY CASE, BEN DECIDED TO LOOK HIS FRIEND UP THAT evening.

"Hello Cook, how are you doing? How's the rock knocking going?"

"Good. I find traces here and there, but that is nothing new. How is it at the mine?"

"Alright. I could use more pay and less hours. Good crew to work with, and the foreman is straight."

"I get that. Have yet to meet a working man who thinks he is being paid what he is worth."

"I suppose. Anyway, seriously, how goes your exploring? I heard rumors and your name came up, claims filed and all. Not sure I heard it right, and thought I'd talk to you. What's up with that?"

"Well, I been out looking at rocks, walking the country, tapping with my hammer. Don't really have nothing yet." Cook looked at the ground as he muttered this, sounding as low as dirt being swept into a dustpan.

"Oh? Like I said, that's not what I hear. I hear you're talking up Jim Argyle, and his place over on Raven Hill. At least that is kind of what I heard in the cage today. Coming out of the shaft, lots of good information gets passed! Don't remember who, but someone said, 'He's found color over there and filed a claim on Raven, at least that is what Cook Eisner told me.' I heard that today. Come on, Cook. He can't have a good claim over on Raven. Most of that mountain is taken."

Cooker Eisner gazed thoughtfully out over the town. Again he muttered, so low Ben had to lean in to hear.

"Yeah, but most people don't think about that. They just hear he has color. And then they go look there."

"True. A lot of folks don't look past their noses. They just hear the rumors and salivate."

"Can you hold a secret, Ben?" Cook said this clearly and a little louder.

"Yeah. I promise not to tell anyone on the cage tomorrow. Unless someone buys me a beer."

"Serious, Ben. I need you to promise. I'm not kidding around."

Ben looked him in the eye and held his hand up as if taking a courtroom oath. "Yes, of course, you ask me to keep it to myself, I do and I will."

Cook mumbled something, too low and garbled to be understood.

"What? I can't understand you, Cooker Eisner. What did you say?"

He looked around to be sure no one was eavesdropping. Then, louder, clearly: "I've found some color. Filed the claim today. It belongs to me. Now I need help to develop it."

Ben half expected something like this, and frankly was a little skeptical.

"Oh? Where?"

Cook glanced warily around again. "Promise me again you will tell no one—not a soul, not even your mama if she comes to town."

"Take it easy, Eisner, and keep my family out of this. I said I'd keep the secret and I will. Now, what gives?"

Cook continued as if Ben hadn't interjected.

"And give me one hundred dollars, or stake me to that much in groceries. Then I tell you where, what, and give you a one third interest. I don't want to go public, I want to keep it between friends. Last thing I want is people bidding in the curbstone market."

The curbstone market was seen in Cripple's wild early days. It didn't occur much now but occasionally it came back to life. The name came from when 'brokers' would bid up mine share prices, standing on sidewalks and hawking to one and all. Usually the winners were the brokers not the bidders or sellers. This was just one of the reasons Cook wanted to avoid it.

Ben stared, not able to get words out. A hundred bucks? That was a fair chunk of change. But Cook knew his stuff, and wouldn't ask unless he had a serious prospect...

"Cook, I could do that. But give me something. I can't put out that kind of money on a four word sentence—'I've found some color.' You have to give me more—where? How? What makes you think it will produce?

"Can't tell you specifics. But let's say it is not Raven Hill or even close. And I can tell you that the ore I got out assays at several ounces per ton. From a reputable assay office, not some backwater chemist working out of a suitcase."

"Several ounces per ton, huh? Sounds pretty good."

"Damn straight. Now, that's all the information you get. More than that will have to wait for your hundred smackers. If not you, I'll approach someone else, but they won't get any more dope up front than I gave you. Remember, I talked to you first, and I hope you and I join forces. What do you say?"

Ben knew Eisner was not a flimflammer. The man had been looking for a mine for quite a while. Never had he found a prospect good enough to approach a partner. This was probably the real deal. Ben took a deep breath.

"Alright, Cook. Done. I don't have a check and sure don't carry that kind of cash. Will an IOU and a handshake do?" He wrote out the note which Cook accepted, smiling. He spilled the where, how, the prospects, and his plans. Their plans now, not his.

"I call it the Emma May. Over on Strombo Point. I got a seam, looks like it may go deep. You will not regret this. Put it there, partner."

They shook hands, and talked the matter over. It was late when Ben left.

Ben said nothing to anyone about this. Staking other miners was a common practice. No doubt half the miners going down the hole every day had some interest in a mine somewhere else. Boasting or gabbing was tiresome and of no use. The stake would pay off or it wouldn't. Either way, there was nothing to be gained by talking about it down in the mine.

This was Cripple Creek, after all. Most every man and woman in the district had aspirations to be owner or part owner of a producing mine. Of course the hope was for a big producer, a million dollar mine. Some would end up that way...

FOR BEN AND HIS CREWMATES, LIFE WENT ON. THE NEXT weeks and months were much the same as the previous weeks and months. By and large Ben acted and really was just a typical hourly wage miner. But in stray moments and especially after hours, he thought of the mine, his mine. Not the hole down which he spent his days but the deepening shaft over on Strombo Point.

He didn't breathe a word to Owens or Juni or Shuler or anyone. At work, he just listened to the gossip, watched, waited. He did talk with Cook often. Now and again, more often as the days passed, he would visit the Emma May. Always he was careful not to be obvious it. He would walk different patterns and be unpredictable. Not that anyone was trailing him or even watching, he was pretty sure, but still.

If the Emma May came through things would change. He and Cook might, if things went very well, just might have to hire armed guards. There was the possibility of serious money here. At first he was sure the smart thing to do was to stay on as a working miner, riding the cage daily. But as time went on, he more and more hoped Cook's color really was a vein, a big wide long vein.

SEVERAL WEEKS LATER, ON THE CAGE, HIS PALS WERE GOSSIPING as usual.

"Did you hear? Old Cook Eisner found color over on Strombo Point."

"Strombo? I thought he was working over on the other side."

"Me too, but I guess not, or he changed. Anyway, he's at Strombo. Filed a claim and all. Named it the Emma May. I guess that is his girl back in Iowa or maybe it was Illinois. You know, the old story: he'd come out and get established, then send for her."

"Yeah, I have a girl in Wisconsin and am still saving to bring her out."

"Well, hell, man, you need to stop spending money buying drinks at the clubs."

"But I'm thirsty after work! Anyway, I heard Eisner got staked by someone. Sounds like maybe he's on to something. Old Cook is digging as we speak. Ain't he a lucky so and so!"

"Well if it goes, Emma May won't be his girl waiting in Iowa, she'll be his wife newly come to Cripple. Anyway, I say, yeah, good for him. Hope it works out. Strombo Point. Who knew?"

The conversation shifted to the day's events.

"Speaking of work, did you see Blake down in the mine this morning? How he took forever to muck just one carful? Is he alright? He looked pale, and gasped some. Had to take a break. He told me it was the brown bottle flu. He's had a spell of that lately. I wonder. When I stop and think on it, I realize he has been kind of ashy and pansy the last week or two."

"My God, I hope it isn't miner's lung. Is it? That's too bad—he was a good worker. Sorry to hear it."

One of the men, a union sympathizer jumped in.

"You know, that is old thinking. Used to be, showing physical weakness was dishonorable and weak. Especially miners. Thing is, all of us miners suffer some from the rock dust and stale air underground. Some take it better than others. But it is a disease, or can cause disease. Just because you are ill, you aren't necessarily weak and unfit. There's no more shame in miner's lung than there is in catching a cold."

"Hah! Any man who can't take a little dust, who is too feeble to get along down there, is a weakling. I for one don't want much to do with a man like that! I hope old Blake is alright and I wish him well, but if he has miner's lung, I am done with him."

The union man disagreed. "The day will come when the government will pay for treatment for miners with the lung problem. You watch and see. I tell you, it is an illness, not weakness."

"Government? Treatment? Never happen here. This is America!"

Mister Blake had just been written off.

Ben didn't hear the chatter about his crew mate. He tuned the voices out. Now he was pondering the street talk about the Emma May. Maybe he was on the road to being rich! After work he went out to the Emma May. In light of the talk, he wanted to see the mine.

"Hello, Cook. You and the Emma May have hit the rumor mill. How is our mine progressing?"

Grinning, Cook held a silver gray rock. "I was planning to come see you this evening. We have some things to discuss!"

He hefted the rock then handed it to Ben.

"This is sylvanite. From the work face. This baby, well, I guess technically this baby's cousin, got us a good assay. Better than I told you when we shook hands. And the mine is full of it! I believe we have something here. We're down forty feet and the vein seems as wide as ever."

They grinned like mimes and shook hands. After a moment Ben broke the silence.

"Maybe it is time to quit my job in the mine and come out here. What do you think?"

"Yes. We need to ramp this up. Your experience will get us going that much faster. When can you start? I need to add a man and we need to put up a sturdy shaft house."

"Give me a week. I want to be fair to my crew mates and the foreman."

"Well, alright. One week. I'll see you tomorrow evening, right? To sketch out plans for the shaft house?"

LON BOSINI WAS IN HIS ELEMENT. HE HAD BEEN IN CRIPPLE for several months. His days were spent in the outdoors. He was doing what countless others had done. He and Cook Eisner were two of thousands of would be gold tycoons. They, he, spent time out in the field, prospecting.

That meant reading the lay of the land, analyzing rock formations, looking for 'color.' Color was oxidation on rocks, or sometimes a quartz vein or some other anomaly. Such might indicate minerals of value nearby. The whole process was seat of the pants navigation coupled with geological knowledge blessed by a sense of rocks and what was in them. A dose of luck certainly didn't hurt.

He had yet to file a claim. In fact he was still getting to know the area. Of course he wasn't working in any mine. This was a typical day, with long hours but fun and interesting. Over dinner with his sister he was telling her about it.

"Up and over hills, knocking rocks, reading the country. Looking for traces others have missed or misread. I love it. And I think I have a lead, a possible point of mineralization. No point in details, it may be nothing. Who knows? Like most miners, I am hopeful."

She smiled thinly, and he realized he had monopolized the evening.

"But enough rock talk, Abby. How is school treating you?"

"Alright. Class preparation and so forth. The children, most of them, are learning well. It is fun to watch them blossom—when one of them suddenly understands it is fulfilling. Doesn't matter if it is long division, noun/verb, or how to sound out words. If the light comes on for them it illuminates the room for me. Plus my meetings keep me busy."

He chuckled although it wasn't funny. "You're a good teacher and I'm sure most of your students do well. They're lucky to have you."

The conversation ebbed.

Lon asked, "By meetings I assume you mean WFM, the mine union. Are you still consorting with the union men? Going to meetings at the union hall? Meet with that Shuler fellow you introduced me to? And how are they doing over there? Any plans to blow up a train or some such?"

"That's not funny, Lon. Please get that tone out of your voice. They don't do such things. These are honorable, hard working men. They just want to improve their lives and also the whole community."

"If you say so, Ab. I just hope they and the CCMOA talk, actually talk and listen to one another. That doesn't often happen. Even if people agree and like each other they don't always listen. So I guess my hope that two antagonists do so is, well, naïve."

"CCMOA? Oh, Cripple Creek Mine Owners Association. The rich men who are used to calling the shots. Of course. Yes, I hope they listen when the men of the WFM speak. I have to say, I don't agree with everything they say they want. But some if that is just negotiating, putting out throwaway items to seem reasonable in the give and take. Still, I hope the MOA listens to the WFM. And, I suppose, vice versa."

"Such things are important, but it is all out of our control. At least you and I are content with our lives. It seems you are happy teaching. And I am happy to be out walking over mountains, finding possible claims. Speaking of getting happy, don't you teach with other young women? I would like to meet your coworkers some time. What do you think, can you introduce me?"

"Well, sure, at the right time. She is busier with school than I am, but I will mention that you'd like to meet her. Don't get your hopes up, brother. There is no shortage of nice men here in Cripple. No shortage of men at all. I say that just so you know, there is plenty of competition."

"Sister Abby, I know that. There is probably more competition in this gold camp for nice women than there is for nice claims! That is one reason I act overprotective sometimes, whether you like it or need it or not."

"I know. You mean no harm."

"Still, I'd like to meet your coworker some time. It'd be nice to see who you spend your days alongside."

He stood and yawned. "Thank you for dinner, Sis. I am done in. I have a lead—the one I mentioned—that I want to check out tomorrow. It is up past Strombo Point. Keep that under you hat, I don't need to tell you. Maybe it is something, maybe not, but let's keep it here between the two of us, alright?"

"Of course. My students wouldn't be interested anyway! Good night, Lon."

THE NEXT MORNING OPENED FAIR. COOL BUT CLEAR. AS THE sun came up the air warmed. Lon walked up a hill, knocking rocks as he went. He was disappointed that the lead he told Abby about didn't work out. Thinking on that, he didn't notice, walked right by, a corner marker. So intent on the surface, he simply did not see the cairn. It was a stack of rocks, obviously man made, standing maybe two feet high. Had he stumbled over it, he would have known someone had filed and proclaimed a mining claim and he should tread carefully. He didn't since he was lost in a rocky self induced fog.

"Hold it right there." Lon found the man's tone annoying and looked up, ready to snarl a response. The shotgun pointed his way stuffed the comment back down his throat.

Hands in front as if to fend off buckshot, he stopped and slowly backed away.

"Hey, I'm just out looking around. I want no trouble."

"Well you sure have a funny way of showing it. Jumping our claim—I could shoot you and there ain't a jury in the county'd find me guilty. Don't you see our claim corner?" Ben nodded towards the cairn, ten feet behind Lon.

"You're lucky today. My trigger finger isn't itching. Just you go look somewhere else."

"Sorry, I just plain didn't see it. Should have, but I was focusing on the ground. My apologies. I will get off your claim right now. You can lower the gun—see, I'm leaving."

Lon turned and walked past the cairn, hands out and visible. He stopped, turned, keeping his hands out.

"Again, sorry. You'll get no trouble from me."

Ben waved the gun then lowered it. "Get lost, you damned pile of dog vomit."

Lon nodded, turned. He forced a few deep breaths to calm himself. Wanting to get out of pellet range, he walked quickly down and away. He decided to check the records to see who had a claim filed on that side of Strombo. Someone as protective as that guy must have something valuable to protect. He'd have to think on that. For that matter, he continued to closely look at rocks as he went. This time he kept an eye out for claim corners.

UP AND BEHIND HIM, BEN WENT BACK OVER THE HILL. COOK noted that he held the shotgun carefully, aimed at the ground.

"I just ran off some damned guy snooping around. He was actually just past the corner marker when I stopped him. I called him a pile of dog vomit. Actually I hoped that'd get him mad enough to make a move so I could plug him. He didn't, just apologized again and left."

Cook guffawed. "Dog vomit, huh? Boy, you really know how to hurt a guy, don't you?! So, what do you think, did he see anything?"

"Don't think so. He took off with his tail between his legs." Ben smiled at that play on words; Cook grinned too.

"Naw, he didn't get to the top of the hill to look over. He just took off straight down. Like I said, if he had stepped closer I might have emptied a barrel his way, over his head. If that didn't do the job...Glad I didn't have to pull the trigger. He left with just a warning from me. I'm sure he saw nothing and won't be back."

Cook grimaced. "Yeah, well, once a mine starts to pay the cockroaches come out. Like you said, we're in the rumor mill now. People—the ones we want to and some we don't want to—are paying attention. We need to be extra watchful. Thanks, good job."

He grinned, muttering 'dog vomit!'

They turned to, moving ore to a burro cart to be taken to the mill.

"Is this stuff still assaying at high gold content, Cook?" Ben picked up a piece. It looked much the same as the sample Cook showed him weeks before.

"The content is jumping around a little, which is concerning but not unheard of. Still, it pays very well. We're good, don't you worry."

He stopped and looked for a long moment at the shaft structure of the Emma May, smiled, and spoke. "You and I will be building our mansions in Colorado Springs before you know it."

Ben thought that response was kind of over the top. Mansion? They had just started. He wondered for a split second if everything was on track. He figured Cook would let him know of any snags or issues. Anyway, he had other things to think about. With a shrug and a nod he shouldered the double barrel and again went to look over the hill. He didn't want to find any piles of goop left by a dog; he grinned at the mental picture.

The trespasser was nowhere to be seen. Ben scanned the whole area. He looked side to side, one hundred eighty degrees worth, and down to the lip of the hill. Amusing images aside, he was relieved to see nothing but rocks, grass and brush.

Lon too was relieved. The barrel end of that shotgun sure did look big and black and worse, the guy holding it looked twitchy. The man was understandably not friendly or reasonable. At least, he thought, the guy didn't put a barrelful of buckshot over my head as a warning. Shotguns pointed his way were not to be taken lightly.

He kept his hands out, swinging his arms in a fast walk as he quickly cleared the area. Being extra sure he was well past the claim marker and over the lip of the hill, he slowed. Looking around, he made sure the area was clean before slowing more. There were no markers, no claims, no people, nothing. Lon felt comfortable in slowing more in order to intently look down.

It was important to examine and understand the ground, its lay, the outcrops, and vegetation patterns. That was one of the subtleties he picked up at the School of Mines—vegetation is an indicator of the underlying soil's mineral content. Picking up some rocks, kicking at others, he worked away down and northerly.

One rock broke off under his heel and the look of the chip stopped him. He bent down, picked it up, and gripped it hard. Turning it over and around, he smiled. The rock sure looked like sylvanite, a bearing ore. Bearing valuable minerals, that is. Everyone in the district sought this. And here it was!

Lon looked around; no one was to be seen. Especially not the guy with the shotgun, he was glad to say! He threw down a red kerchief to mark the spot and paced back fifty or so steps. Looking around, he eyed the rocks and the vegetation, satisfying himself he had the right spot. Then he picked up some rocks and made a cairn. He reached in a pocket and got some paper. Just for a time like this, he carried several sheets of paper pre-filled out with his name and 'I stake this claim' on it. He filled in the date and location, folded it up and put it under a rock on top of the cairn.

Lon looked around again. He memorized the small landmarks to be able to come back to this precise spot. As he did this he confirmed that he still had this side of the hill to himself. Then he strode off.

It was about this time that Ben peered over the top of the hill. He saw nothing, and was relieved that the trespasser had indeed walked away. Just as well, he didn't want trouble any more than the interloper did. Ben grinned once more about the encounter, turned and went back to the Emma May.

AN HOUR OR SO LATER, LON WALKED TO THE LAND OFFICE. THE pace he set was purposeful but not too fast. Last thing he wanted was to draw attention so he tried to blend in on the street. After he entered the building, he pulled out the papers to verify the claim location, date, and so on for the actual filing. On the way down he decided on the name: the Double I, for his and Abby's Irish and Italian heritage. Papers filed and with the proper receipt and acknowledgment, he set out again.

Back up the hill he ambled, again not too fast. Quickly he built three more cairns. They were about two feet high and clearly man made. The purpose was to mark the corners of the claim. He made sure they would stand out more than the one he bypassed only to run into the shotgun. That done, he filled a gunny sack about a third full with rocks. That was all he could carry anyway. The rocks he picked were ones he was pretty sure were the good ore. To an assay office he and his cargo went.

"Hello. I need an assay on these samples." He dropped the sack at the feet of the chemist. The thump echoed and several other workers looked up.

The chemist motioned for the sack; Lon drug it towards him and handed it off. The man hefted it and thought a moment. "Alright. You want the estimate of gold, silver, copper, lead and nickel per ton, right? It'll cost you seven dollars an hour of lab time, plus materials. Need a twenty dollar deposit and a name. Where'd you get 'em?"

"Up on the hill." Lon had no intention of telling the location and the assayer really didn't expect him to. "Name, call it the Double I. My name is Bosini, Lon Bosini. That is b-o-s-i-n-i. Here's twenty. When can you have the results?"

The chemist handed him a receipt. "Check back tomorrow after about three."

Lon kept this all to himself, telling not even his sister. The next day he dug an assay pit, a hole six feet across and four deep. The samples he collected there went to another assay office. Lon wanted to hedge his bets, make sure he was getting a true picture of the potential. After both reports came back favorably, he allowed himself to think maybe he had something. Enough anyway to go on. He decided to share some news with her at dinner.

"So how my big brother the prospector doing?" Abby smiled. She expected the ready answer telling of a lot of looking but precious little finding, and a scolding for calling him big brother.

"Well, little sister, it is like this." He knew she didn't like that but after all she started it. Family theatrics aside, he made a show of moping. And he looked down as he moved his coffee cup around like he had bad news to tell. Then a smiling look at her. "I found color."

As she knew, finding color meant maybe there were valuables within. It was the phrase everyone used to announce good news, progress, and possible profit.

"Well? Are you going to tell me about it?"

"Not only did I find color, it was good enough for me to file a claim. Up near Strombo Point. And, it assays out very profitable. If the vein is true and I filed the claim over the right spot, things will go well. I named it the Double I, for our heritage, Dad's Irish and Mom's Italian forebears. It looks very promising, Abby. I hope to make a pile of money!"

"Lon! That is wonderful!" The excitement in her voice caused some patrons to look up. She lowered her voice and asked, "Is there anything you need me to do?"

Lon glanced around, a little startled by her enthusiasm. He spoke quietly.

"Not now. Just keep it to yourself. Oh, and you own part of this mine. I took the liberty of putting your name on the claim as well. Ownership, me fifty point one percent, you forty nine point nine. Worst that can happen is you have to pay some tax if we make money."

"Oh. You didn't have to do that."

"I know I didn't and it is my pleasure. After all, you are my sister and you have graciously fed and entertained me often. Anyway, to your question: Don't do anything, don't talk about it. It is important that nobody hears about it. If so, they'll want to get over there and interfere, file cross claims, claim prior work, something to make it hard for us."

"I understand and no one will hear it from me."

"Now, I have enough money to get development started. The mine should—I hope—pay for itself shortly. If anyone asks, tell 'em I've been looking over on the other side of town, general like. No one will ask, I imagine, but if they do, send them the other way. And this will all work out real well, sis."

"That is good news, and like I say, I'll keep it under my bonnet. And Lon? Congratulations!"

He smiled, nodding. "Now, not to change the subject, but how is school going? Anything new in the 'ABC' business? Oh, and that reminds me, I'm still waiting to meet your friend. She must be busy.

"Well, she is seeing someone now. They seem to be getting serious, so I can't really try to break that up. Sorry, if that changes I'll tell you.

"I understand." Lon had been hopeful but was not surprised an eligible, educated woman was spoken for. "Oh well. Say, you aren't still talking to those union men are you?"

"What if I am? None of your business, mister not my boss!"

It was indeed his concern, but he stuffed saying that, rather just laughed. "No, I guess not. Didn't mean to pry. I'm just interested. All I ask, Abby, is be careful."

He put on a wry smile. "That big brotherly advice will have to suffice, although I know you are careful anyway. It makes me feel better to say it. Now, I have to get going. Good to talk with you and I'll keep you up on developments."

IN FACT, ABBY WAS CAREFUL ABOUT WHERE SHE WENT AND who she told about it. She said nothing as he finally left, but she was headed to a union get together after she saw Lon off. She wasn't about to tell him that—he would have tried to stop her, or worse, accompany her. Actually his brother instincts made him insist that he see her safely to her room.

She was used to his being a little protective. Rather than fight it, she just let him walk her home. After he left she would go right out on her own to the meeting. With a number of big beefy friends there, most of them as concerned for her as her brother was, she knew she was safe.

ABOUT THE TIME LON WAS TELLING ABBY ABOUT THE DOUBLE I, Ben sat down heavily. He was in Cripple, at a corner, on a bench in front of a sporting house. He mused moodily, wallowing sourly in thought. He had no interest or energy for anything today, certainly not playing with the girls.

His outlook gyrated between philosophical and angry. Like so many, he thought he had the tiger by the tail when really the tiger was toying with him. This happened most anywhere, but especially in a town like Cripple Creek.

He had agonized on his future after putting up Cooker's stake. He listened to all sides of the argument, mostly silent thoughts and viewpoints in his head. Against the advice of one small voice, he quit his mine job and borrowed on his house. After all, he needed to help out at the mine. So he cut work ties, said goodbye to his crew mates, and forged ahead.

The loan he took out was used to develop the Emma May. He was sure as could be that it would go big and he could repay the loan easily. It was just a matter of digging and processing the ore. Another small voice had told him he would soon be rich, able to go hobnobbing with Stratton.

The thought of Stratton sidetracked him for a minute. It was a Cripple Creek legend. Winfield Scott Stratton drove the story. He was a carpenter from Indiana. Cornfields held few charms and he came west. As a prospector he was largely self-taught. He did take some course work at the Colorado School of Mines. Over the years he walked and examined rocks all over western Colorado. Silverton and other areas as well as Cripple Creek got examined. Winters, he retreated to ply his carpentry trade in Colorado Springs. Summers the saw and hammer were stowed, and his first love got all the attention: he roamed and knocked rocks. This went on for years. At last, after much inspection and thinking, he staked a claim near Victor. It was July 4, 1891. Against the advice of many, he held on and developed it. That patch of ground became the Independence Mine. The big strike was three years later, in '94. Winfield Scott Stratton became many times over a millionaire, the first in the district. Several years later the man moved to Colorado Springs. He lived in a fairly modest home there. Not ostentatious, he was quiet and fairly solitary about it. No splashy mansion, travel, or social whirl for him. But he had money oh yes he did.

Most everyone who could swing a rock hammer in the Cripple Creek area was sure they would be the next Stratton. This brought Ben back to the Emma May, the mine of his dreams. For quite a while, the vein was wide and its ore rich. Things looked better every day. Maybe they weren't crazy to think of a place in Little London.

Then the vein pinched out.

Cook and Ben were perplexed, stymied. No matter which direction they looked, no vein. No ore worth the effort. No riches. No mansion in Colorado Springs. For Ben, no way to pay off the loan on his house. It looked like he'd have to go back to sleeping in some rooming house. The two of them would have to go back to the mines at three dollars a day. If they could get on somewhere.

But they kept the claim and despairingly hammered off a sample or two every day. The hope was they missed the turn of the vein. It would take just one swing of the hammer to turn things around. Stranger things had happened.

Right now, sitting on a bench on the main street, Ben didn't give a damn about much. Certainly he didn't concern himself with labor and the union and owners' arrogance and all that stuff. Angry at the world and at himself, he almost enjoyed wallowing in the misery. Hell, he realized, he and thousands of others had just missed out on becoming rich. The western US was full of miners who thought they had it made and then saw the paying ore peter out.

Finally and once again, he told himself to go start to rebuild. The Emma May was likely history. Let her go and pay attention to the important things, like working conditions for all miners. Suddenly, he decided to go to the Union meeting. What the heck. Get back in the fight, right?

Ben sat up, smiled, and counted blessings. Of course health and friends were high on the list. Colorado blue skies and sunshine. And living in Cripple Creek! This was the greatest gold camp, the most exciting spot on earth. No better place to be right now. Sure, he was down at this moment, but any rock out there could be the clue to the next big strike. And it could be his!

BEFORE HE WENT TO THE UNION HALL, HE THOUGHT BACK. HE recalled a talk with his big friend Juni, about how extraordinary life was here at ten thousand feet.

"Man, have you heard about the strike over in Elkton? You know, the little settlement up the road? Some guy hit a vein with fifteen, twenty ounces per ton! And he was on his last dime at the time I guess. Knocked one last rock, and boom!"

"Yeah, I heard about a super rich strike in that area. Wow, I wonder how that felt. You know, Juni, what fascinates me, besides ordinary Joes hitting it big, is us."

"Us?" The big man looked at Ben like he had sprouted a tail.

"Yeah, I mean regular folks going around, not only getting rich like I say. But the sheer excitement of life here— mine strikes, miners coming and going twenty four hours a day. Think of Bennett Avenue, the main business street, running across a hill so steep that the two lanes are on different levels. And then the sporting houses on Myers Avenue just next to that. Then there are all the fancy hotels catering to rich mine owners. Those cats are calling the governor for favors and you know the governor asks them for favors too. Don't forget all the newspapers with their headlines, and the new houses and mansions being built, and the Vice President candidate coming to town..."

"Yeah I was there you know. That Teddy Roosevelt, he got himself in a serious scrape over in Victor. That would have been bad, real bad, if the future Vice President got beat up in our camp. Talk about headlines! Yah, this is quite a place, it is."

This was news to Ben. He knew that Teddy was forced to put up his dukes there, but had no idea his friend had witnessed it. "No, Juni I did not know you were there. I bet that was something. What were you doing in Victor? Talk to me. I read the papers but tell me what you saw and did."

"Well. I was over there talking with some friends about my..." He grinned, sheepish and sly. "Doing a little work around my after hours sideline." He mimed slipping a rock into his pocket, and winked.

"Anyway, I was over there and someone said Teddy Roosevelt was coming to town. I thought, fine, a politician, not a big thing. Then I remembered, he might be the Vice President, maybe some day the President. He had a reputation as a good speaker, so I figured he was worth going to see. Boy oh boy, was that show worth seeing!"

"I guess some of those guys in Victor didn't like him."

"You got that right. I guess McKinley, the Presidential candidate, was against the silver standard, in a big way. He pushed through the bill repealing it, which caused the silver crash of '93. And the Panic of '93 followed, and we were all out of jobs, and railroads folded, and it was a hellish mess all over."

"How well I remember."

Juni continued. "And many in Victor remembered that Silver Act. They thought, and I have to agree, it was a betrayal. And they hated McKinley for it. Far as they were concerned, young VP candidate Teddy was just as bad. Old money, big business, young guy who did McKinley's bidding is what they figured. They wound themselves up as Victorites can do, know what I mean?"

Juni didn't wait for Ben to respond, getting into his story.

"The guys in the crowd as he spoke were getting worked up. Saying—loudly saying—things like 'We'll show that politician a thing or two,' and 'Let's send a message to Washington' and other even less friendly ideas."

"So the crowd was laying for him." Ben found this fascinating.

"Yup. At least enough of them to make trouble. So, he—TR—didn't help himself. He said things like he approved of Governor Peabody's anti labor stance. Gee, Peabody and the mine owners were like a miner and a dancehall girl at two in the morning, you know? He was hated in Victor."

He made a gesture which left little to Ben's imagination. He grinned and nodded.

"And—get this! Talk about ol' Teddy saying the wrong thing at the wrong time! He said, actually came right out and said, that those who liked the silver standard were 'crackpots and impractical visionaries.' Crackpots! Impractical visionaries! Boy oh boy, how's that for big fancy words? Words that really riled up a bunch of miners who lost silver mining jobs. This was a blue blood New Yorker lecturing us, for God's sake!"

"Easy, Juni. Politicians and hot air and all. You can't take 'em too seriously. So, what happened when he got to town?"

"Maybe so, but really, talk about a match in a dynamite factory! I mean, really, a man on the national stage ought to know better. He should not go out of his way to antagonize his audience. Which ol' Teddy sure did, on purpose or not."

"I doubt it was on purpose, Juni. It was probably just east coast arrogance."

"Anyway, then he headed over towards the Armory there in Victor. A crowd followed him, threatening to tar and feather him, run him out of town. Inside the Armory building people were just as rowdy. Roosevelt tried to speak a few times and finally gave up; the crowd wouldn't go quiet enough for him to be heard. He and the mayor snuck out a side door and were headed back to the train."

"Wow, they got Teddy Roosevelt to retreat. That is a rare thing. It must have been something!"

"And it ain't over. They left the Armory and the crowd followed. They caught up to TR, throwing rocks and yelling. His hat got knocked off and his glasses too."

"Well, miners like to skewer stuffed shirts. I guess he kind of asked for the hot reception, no?"

"And it got hotter. The crowd started to close in—I was on the fringes and it was starting to look ugly. Suddenly, though, a big guy stormed up, grabbed a 2X4, and swung it. He cleared a path and TR followed him to the train. I really think he saved our future Vice President from a beating or worse."

"Wow. What, did TR do then? Leave the county? Go back to Colorado Springs and sip tea with the millionaires?"

Juni laughed.

"No. You gotta love the guy. Early in the day, he almost got thrashed, not one on one but by a mob which would make anyone think twice. But he didn't run away with his tail between his legs."

Here Ben had a sudden image of his mine intruder months back, and suppressed a laugh. Juni went on.

"No, ol' Teddy stayed in the area, finished his trip. After Victor he went right over to Cripple Creek. There he followed the plan and toured a mine or two. That wasn't enough—after a dinner he gave a speech, a real barn burner I guess."

"Yeah, Teddy Roosevelt can talk a good one."

Juni nodded.

"And after that speech he went out of his way to talk about the day. He gave credit by name to the big guy who saved him. It was the postmaster here, a guy named Danny something, can't remember his name, but he's an Irishman. Sullivan, that's it. Danny Sullivan. Roosevelt actually said the guy saved his life. I guess the crowd loved him, not for his silver views for sure. But he stood up to the bullies and he gave credit where it was deserved. He got huge applause, I guess."

Ben grinned. "Wow. That is quite a tale, Juni. Only in Cripple Creek would a candidate for national office get taunted with tar and feathering. And to make it better, nowhere else would the candidate smile and shake his fist at the taunters! I'm not surprised the Cripple Creekers ate it up! We love a good story here."

Juni left, with appropriate talk of getting together again soon.

As Ben looked back on that conversation, he was struck. Even though his personal situation had just gone down the tubes, he still was happy and optimistic. He thought, 'God in heaven, but I do love this place!'

Then his mood turned and he spent a few more minutes fretting over being pretty much broke, and grieving the Emma May. The mood turned again as he stood and faced the music. Slowly, wistfully he stood and sauntered towards the union meeting.

VI

FOR A WHILE, THE EMMA MAY TOOK OVER BEN'S LIFE. SEVERAL months, maybe even a year, had gone by since he had gone to a union meeting. Trying to make his own mine pay was more than enough for him. He didn't have time or interest to look after the affairs of miners he didn't know and would never meet. He was not current on developments the miners faced day to day. He had pretty much lost touch with union issues. Now it was time to try and get back into the mix.

The last meeting he had gone to ran through his mind. Then, he remembered, Big Bill Haywood wanted to change procedures at the local level. As things stood then, all decisions were voted on by the members. Local officers acted at the direction and approval of the membership. Haywood viewed that as unwieldy and slow.

He was the Executive Secretary of the national union, the Western Federation of Miners. As part of that job, he wanted to adjust those procedures. The plan he offered cut the membership direction down, almost out entirely. He wanted authority for the local Union officers to act on behalf of and without input from the members. That authority included the right to call a strike or slowdown.

As Ben recalled, the members, at least those at the meeting, were wary. Most of them didn't want to give up final say-so. They were leery of ceding too much power to the union and its officers. Haywood was not a low key, let's wait and see type of leader. He was all in for the union and doing it (whatever 'it' happened to be) now. Some feared he would lead them out onto thin ice.

After that meeting, Ben recalled that he had talked it over with his friend Owens. It had been a while and the details were fuzzy. He remembered that Owens thought it a good idea.

"Heck yes, the members don't need to vote on every little thing. That is what officers are for—to run the damned place. We elect those officers, don't we?" He waved his hands and swiveled his head as he usually did when putting out his ideas. "We should put good people in as officers and we have. I for one trust Bill Haywood and Frank Shuler and others not to steer us wrong."

"But Owe, it isn't voting on only little things he wants to cut out. I agree, who cares if the officers spend three dollars on a dinner, or stay in a fancy hotel while in Denver on business. But everyone cares if, say the officers suddenly call a strike or protest. For no explainable reason. That concerns me and a lot of other men. After all, it is members' lives they are playing with."

"The officers wouldn't send us out for no good reason. And they would certainly be sure to have member support before they did." He said this as fervently as a Southern Baptist calling dancing the work of the devil.

"I don't know, Owe. What can we do if they try to run us out on strike without our support?"

"But they wouldn't do that, is what I'm saying."

Ben snorted. "Your faith is touching. I tell you, this man Haywood hates mine owners. And more worrying, he hates how things are set up, with private ownership of property. I guess he believes we should all share everything or something. Not sure how that would work. Anyway, hate isn't a good basis for life. This man seems so blinded by it that he'd do anything to hurt the owners, miners be damned. I really question if he should have the say he wants."

Owens' hands were windmilling again. "But Ben, he doesn't want that say for himself, what he wants is for the officers of the locals to have the say. The men on the ground, so to speak."

He grinned, and stood still for a moment. "I guess they're really the men under the ground not on it, but I really think giving this to him, to the officers, will be good for the miners and for the WFM, and not so good for the owners, you watch and see."

As Ben climbed the stairs toward the meeting room he mulled that long ago conversation. The room was pretty full from the sound of talking coming from it, the volume growing as he neared the door. Being a mine owner for a while hadn't changed his reservations about the proposed changes. He looked forward to seeing how things had progressed.

Since he decided to attend on the spur of the moment, he arrived a tad late. A convenient seat was in the back, near the end of the row. He settled in, listening with half an ear. He still thought on Owen's faith and trust in the union leaders. In the background, Shuler was going on something about the mine owners and the Cripple Creek Merchants Association or some such. As he spoke, Haywood sat behind him, glowering.

It was the same song, another verse. Same old, same old. Stuck in the same ruts, is what he heard. Ben almost got up and left as that realization came to him.

But he didn't. He stayed and sat, but drifted off. This time it wasn't union procedures he pondered. He was trying to come up with ways to make the Emma May pay. Or at least how to get some of what he had put into it, back out.

Several scenarios came to mind. Maybe he and Cook could sell it. He wondered what happened to the rube that he ran off a while back. If Ben could only find him, maybe he could be persuaded to buy it. Was there a claim nearby which it could be consolidated with? Maybe some rookie, a new-comer to camp, could be hornswoggled into buying a share. Or maybe...Hell, he couldn't come up with any viable plan. He turned attention back to the room he was in.

From the sound of things, Shuler was winding up his talk, the subject now not the MOA but newspapers.

"And so we need to be aware what we read, where you get your news. Remember, most of the papers are owned by or are in the pocket of the owners! Read the miners' paper. And the union newsletters. And of course us, talk to me and the other officers. We're all in the know, so ask if you have questions. Don't take anything you hear or see at face value. Especially from the owners' newspapers."

He paused, picked up some pamphlets and gave them to a member in the front row. "I have a stack of the WFM news-letter for the month. Here, take one and pass them on."

"Also, watch your talk, men. Be careful what we let the owners hear. The members, and their friends, should be care-ful what they say about the WFM at all times. Especially so in front of the foremen and superintendents. They are listening and planning and they talk to each other, yes they do. The owners take every chance they get to sell their message. Don't give them ammunition! And tell your coworkers and friends about this. If you have to say something about the WFM, make it positive. Better yet, don't say anything, let them stew and guess! Don't give the owners anything to work with!"

He looked around the room, into each face, making sure his message hit home. "Questions? No? Good. Now, I want to welcome back 'Big Bill' Haywood. He is a busy man. But our Executive Secretary has made time to come down from Denver to talk with us." He stood aside.

Not long before Shuler's oration started Lon and Abby set out. Not that Lon cared about what Shuler did or said. He took his time walking his sister to her home. Once there he seemed reluctant to leave. Maybe, Ab thought, he really did hope to meet her coworker and friend. Or he was feeling lonely. Or wanted to celebrate his Double I mine. Too bad, she didn't have time tonight. She had somewhere to go. No luck for Lon to meet the friend; she was away. Abby fidgeted and stalled and finally Lon left.

Shortly, Abby scooted out and went the other way. What with Lon hanging around, she was a little late arriving at the meeting. As she came in, Shuler was talking and Big Bill Haywood was glaring and fidgeting up on the stage.

The first time she heard Haywood speak she had seen through him. Far as she was concerned he was an angry bully and best avoided. Being busy, she had never really stopped

to think about a man in his position with such a tempera-
ment. Tonight his behavior brought it out. He, Haywood, was
always fuming about something. She was glad she didn't work
with him day to day. Talk about a caged lion! No telling when
or why a temper like his might explode. Glad to be a spectator
only, she slid into a seat in the rear, on the end of the row.

As Abby got herself got settled she listened to Frank
Shuler finish up, something about the miners watching what
they say around owners and management. As if they didn't
already! How stupid did he think his miners were, anyway?
She wondered how long it had been since the man had actually touched a piece of ore or hefted a shovel. It was a marvel
that he had the nerve to lecture like that.

Shuler's talk dribbled to an end and mercifully, he introduced the main event. Big Bill Haywood stood and loomed as
he walked to front and center. He paused dramatically.

Glancing around, Ab made mental note of who was
present and how many were there. Two seats over sat a man
she'd seen once before, a while back. She hadn't seen him at a
meeting for several months now, and thought back to the first
time he had caught her eye.

Then, she found him kind of attractive and was mildly
intrigued. She didn't know who he was or anything about
him. She sure wasn't going seek him out. Now, today, here he
was, an empty chair away. And sure enough, he was peeking
sideways at her, trying not to be obvious.

Everyone expected Haywood to launch into his tirade.
He had a little surprise, a break in the routine.

"Frank has given us plenty to think about. What say
we take a little breather." He pulled a watch out of a pocket
in the front of his blue denim trousers. He peered at the

timepiece closely as if checking the progress of a lit fuse sputtering towards its stick of dynamite. Stowing the watch, he spoke tersely, sounding heated as he blurted, "Meet back here in fifteen minutes." He quickly left the room.

Not that she looked closely at mens' clothes, but Abby had not seen a pocket like the one Haywood stashed his watch in. It was small and above the main front pocket on the right side. It was made for just such a watch, she realized, and thought it a creative idea and useful thing. She mused on how clothes could be practical in little ways like that and then her mind started to float off towards Lon and his news.

Would he hit it big? Or would the Double I stumble along, making money but not a lot of it, like many mines? Or, worst case, it might peter out and become a shaft Lon threw money down. Her brother, she was sure, would have fun whatever the outcome. The fate of the Double I was kind of interesting and she wondered idly if she too would be rich. However things went, she intended to stay and teach her kids, right here in Cripple Creek. That in turn made her think of some of her students who needed some special attention, and she started thinking of ways to attack their problems. Like most thoughts, these flew in and out her head in a flash.

In the background people enjoyed the break. They stood, stretched, and chatted. The talk was in low tones.

A not unpleasant voice intruded. "Who put a burr under his saddle I wonder." The man one seat over smiled as he mentioned Big Bill's running off. "He sure took out of here fast."

Abby offered a little smile. Good to be friendly but not too much.

"Mister Haywood is forever in a hurry. He's always scowling and tensed up. He probably needed to"—here she almost said 'drain his dog' but realized that was crude, even for a miners' union meeting. She was able to say instead, "see a man about a horse." Try as she might, she couldn't help but blush as this came out.

The last thing Ben expected was earthy humor. He laughed in surprise. "You may be right about that. You sure are on the money about his being intense. He gets so wound up sometimes, you're afraid he might explode."

"Yes, Big Bill knows what he wants doesn't he? And he is sure ready to go right after it, whatever the obstacles."

Ben slipped over to the seat next to her. "May I?" This was the redhead with brown eyes he had seen months before. When he moved she didn't flinch nor did she say it was alright. She just smiled coolly.

"I'm Ben. Ben McNall. I think I've seen you here at some of these meetings. Do you work for the Union?"

"Me? No. No I don't work here, not at all. The local school is where I work; they pay me to teach the children. I am Abigail Bosini, by the way." She extended her hand. Ben didn't expect this from a woman and hesitated a tiny second before shaking it.

"Nice to meet you, Abigail."

She continued. "Call me Abby. Yes, you may well have seen me here. I attend these meetings as often as I can. The whole labor-management partnership, you know, wages and conditions and all is fascinating to me. Not just pie in the sky theory. I stand for the working man and woman. I'm all for fair pay and decent working hours."

She looked sharply at him. "Including equal pay for equal work—it is not right or fair that a man teacher with similar experience to mine is paid more than I am, just because I'm a woman. But that's a topic for another day, not really pertinent for a miners' union meeting."

Her face lit with a smile; her eyes, not so much, and she went on.

"Actually, I'm here because I think a workers' union is the way to accomplish some of those goals. This is the only way I can support the idea. I wonder if the union would even accept my money if I tried to join? Anyway, that is an idle thought. Enough about Abby Bosini. And you—which mine do you work in, Ben?"

He was swimming in her brown eyes, the red halo from her hair reflecting in them. It was a job for Ben to force himself to pull back and respond. He wanted to impress this woman. He almost bragged of being part owner of the Emma May. But Miss Emma May was showing herself to be a barren and greedy hag, not a rich generous mistress. Before he spoke he was able to remind himself that many men had tried to open a mine. There was no shame that the vein pinched out.

"I'm, ah, between mines at the moment."

A little disappointed. "Oh, I see. And what..." Her thoughts raced. She was sure—well, she hoped—that there was a good reason for his being unemployed. In Cripple Creek, jobs were plentiful. Anyone not working did so by choice. That choice usually involved drink, drugs, or gambling. Or labor agitation. This man didn't seem the type for any of that. He seemed solid—why else would he be at a union hall? She hoped to learn more. But Big Bill stomped back into the room as she pondered how to respond.

"Alright, men."

During the break people scattered and they slowly started to sit again. Haywood hoped to get folks to sit and listen. He cleared his throat and spoke louder.

"Alright men," he repeated himself. Then he threw an apologetic glance towards Abby and a small group of women over in one corner of the room.

"Sorry, ladies. Let me rephrase that: Alright, folks."

He actually smiled gracefully, then his tense, hurried habits took over.

"Let's get started here." He waited for a few seconds and when people still didn't move to their seats and become quiet, he simply forged on.

"It is always good to come to Cripple. Today I want to talk about the plan I've been working on and we have explained and discussed at several meetings. My plan will put authority to act in the hands of your local officers, to act without the delay of putting issues to a vote of the members."

He also intended to reinforce his efforts to limit what his members read and heard. Last thing he wanted was for WFM men to see the other side's point of view. He wanted to steer the union and the members towards confrontation with the MOA. Best way to do that was to make them the enemy, dark hearted and greedy.

He thought this but didn't pause as he continued to explain the benefits of his plan.

"Like I've said, the local officers need flexibility to act when the mine owners stomp on members' rights. Which they do often, I remind you! And your officers need this. They must be able to act quickly and decisively. When you are at war like we are with the owners, you can't be putting every little issue up to the yea or nay of the members.

Generals don't ask their soldiers for advice, do they? No! We at Denver headquarters have great faith in the officers and you should too. They have common sense and willingness to act in your interest. After all, if we don't take care of our members, why are we here? Are there any questions?"

A member stood. "Mr. Haywood, has this change been made in other locals?"

"When it happens, this change will apply throughout the Federation. Many locals have approved the change and frankly are waiting for the big camps to agree. When I say 'big camps' what I mean, really, is Cripple Creek. This is— you are—the flagship local of the WFM."

He looked around the room to be sure the members knew how important they were, and, convinced they agreed, went on.

"The change is designed, like I have said, to give the locals the means to respond when the owners take some unreasonable action. And that takes place often as you know. It will also let us take the initiative, take action before they take some illegal or inappropriate act."

Another question from the floor: "I don't know, Mr. Haywood. This would be a big change for us. Seems to me that if we go ahead with this, we should also stage new elections at the same time for new officers. I mean, since we are putting new responsibility and authority in the hands of our leaders we may want to adjust our officer ranks. What do you think about that?"

Big Bill Haywood took a good long look at this speaker, visibly memorizing his face. His response was icy. "You trust your officers now, don't you? If you do, why wouldn't you trust them with a little more authority? If you don't trust them maybe you shouldn't be in the WFM." The man sat down, wilted.

Ben and Abby traded a glance at the man's obvious discomfort.

Haywood looked away. "Next question."

Abby and Ben again traded glances as the silence lengthened.

Frank Shuler spoke. "I am glad we have had a frank and full discussion. No more discussion? Alright, we have covered the issue. I will entertain a motion to close discussion and to broaden our local officers' authority per Mr. Haywood's written proposal. You have all seen copies of this, I believe."

Someone said "I so move."

Another, "I second the motion."

After a moment, a third: "I move that this be passed by acclamation and be recorded as a unanimous decision by the membership."

Shuler tried to contain his smile as he glanced at Haywood, then spoke. "All in favor, say Aye."

A chorus of 'ayes' shook the room. Shuler asked for nays and got a few, most shaky and muted. In contrast, the man who asked the last question boomed his 'nay.'

"Motion carries. The record will show this change in procedure as having unanimous approval. There being no other business, the meeting is adjourned." The bang of the gavel sent most members out into the night.

Abby stood and looked at Ben. "You mentioned something about being between mines. Are you involved in two of them? Not to be presumptuous, but this interests me. I'd love to hear more. As I said, during the day I teach, but maybe some evening we could chat."

She flushed a little at being so forward, but looked him in the eye for an answer.

"The Emma May, that's my mine. Yes, that would be nice. Can I stop by the school some day and make arrangements for the evening? The thing is, I have a significant interest in a mine which was going..."

Shuler walked up and grabbed Ben by the elbow, in a friendly but direct way. "Haywood wants to talk to you." As he steered Ben towards the front of the room, he looked back at Abby. "Sorry to cut you out."

Ben too looked back and said, "I'll be in touch, Miss Bosini."

HAYWOOD SHOOK BEN'S HAND. "GOOD TO SEE YOU AGAIN, McNall. You've missed some meetings."

"Yes, well, I have been busy with a mine I have a large piece of."

Haywood eyed him sympathetically. "And I'd bet that the vein went out on you and took all your money. And you will be looking for mine work, which is why you're back."

Ben felt naked, transparent. He nodded.

Haywood smiled. "Don't take it so hard, man. I've seen it before many times—it happens more than you'd think. A man sees a little color and suddenly he's going to be rich. He can call his own shots and he won't need to work in the mines and to hell with the owners and foremen. And then reality bites him in the hind end and here he is again."

He roared with laughter; Shuler joined. Ben smiled weakly.

"Hell, man I understand. I have had this conversation with hundreds of men, good men, good miners. Believe me, life goes on. Right now it seems bleak. Things will get better, they really will."

"I guess. So, you wanted to talk with me?"

"Yes. Ben, I am glad you are back in the fold. The thing is, the WFM needs men like you. Now we have some authority to move. We can take the initiative. You and your friends can look for good things to happen. Good for the miners, good for the Union, good for the mining camps. Maybe not so good for the owners but they will get everything they deserve."

He glanced around to be sure the room was empty. Only Shuler and Ben were with him. He lowered his voice. "The thing is, Ben, some of the locals—local chapters of the WFM—have let themselves get pushed around. That pains me, it really does."

He touched his heart as if feeling a jolt of agony, and went on.

"Now, now we can act. We can force those miserable owners and their merchant lackeys to the negotiating table! No more sitting around waiting for them to drop us some crumbs. We can grab the loaf out of their hands! And, best of all, we can do some things to let them know we are serious and we won't be cowed. Some sudden events to get their attention."

As he said this last, he brought his hands up and suddenly spread them apart, as if exploded. And he winked. Shuler looked away. Haywood continued.

"We'll be hitting them where they hurt—their wallets. See, if we can stop the pipeline somewhere, we can get them to talk, offer better terms. There is more than one way to sit them down and deal."

"Stop the pipeline? We're not plumbers. What do you mean?"

Haywood chuckled. "What I mean is, we stop the production of wealth. That is what this is all about, no? Production, and sharing, of wealth. Especially a fair sharing. Now, by 'we' stop production, I don't mean the local officers. 'We' means me, Bill Haywood, and my team. My circle of friends and advisers. You can be one of that team, Ben. We'll come back to that."

He paused, glancing at Ben then Shuler before going went on. "You're right. We're not plumbers. But think about it, mining is really a matter of flow, of a stream. The owners make wealth out of the ore our miners dig out, for cheap, almost starvation wages. Right?"

He was pacing now. "Ore comes out of the mine, is hauled to the smelter, refined, and the gold is sold to banks or the government. If you want to get everyone's attention, just dam up the flow somewhere. Stop the wealth. See, if we dam up the stream somewhere, anywhere, we get the owners' attention."

Ben nodded; Big Bill hardly noticed since he was onto one of his favorite toots.

"Now, by pipeline, I mean we stop the owners from making wealth out of that ore. Simple and easy. There are several ways to do it and here are the most effective."

Here he held up a fist and stuck up fingers as he went.

"1. We strike the mines. That stops ore from coming out to be sent and refined. 2. We can somehow stop the transport of the ore. That's tough since the mine owners are in cahoots with the wagoneers and railroaders. 3. We can strike the mills, stop the processing of the ore altogether. There are probably other ways too, but like I said those are the three big ones."

Ben nodded thoughtfully, thinking the three options through. Then his mouth ran away with him. He spoke before he thought through giving advice to, joining up with, a hard man like Haywood. He was committed before he truly considered it all. Later he would look back and realize this was a major T in his life road.

"The chokepoint, from what you say, is the mills. There can't be that many of them. A few small ones here, Canon City, and Colorado City, right? Does any of Cripple's ore go to Denver for processing? How are the workers paid? Any better than our miners? How many millmen are there? Are they open to the WFM, or can they be convinced?"

"Good questions, Ben. Questions like that make me see that I need you. I need advice like this. And I need your eyes and ears out there. You work with the men daily, swinging hammer and moving muck. And I admire that." He paused, gathering his thoughts.

"The men—and women—here looked surprised when I said we are at war. Make no mistake, we are in a fight for survival here. Not everyone knows that but I do and Spencer Penrose knows it too. Now, in a war you need information on your side and on the enemy. I need to know what the troops on our side are thinking. Not what they think I want to hear. I need to know what the men think and say when it is just the miners. When no one is around or listening."

He approached, put his hands up on Ben's shoulders and looked him in the eye.

"Pay attention to what the miners care about and talk about. And tell me. If I'm not here talk to Shuler, he'll pass it on. And he needs to know that stuff anyway."

Ben glanced at Shuler, who smiled and nodded even as Haywood continued.

"Also, I need to know whatever you hear from the other side. Can you help us, Ben? Can you help me?"

It was a rush of big events for Ben, all in a day or two. He felt almost overwhelmed.

There was the unexpected demise of the Emma May. That was more than enough, but its failure could, likely would, cause him to lose his house. Bad news. Then, good news: meeting an interesting—and interested—woman at tonight's meeting. Last thing, good news or bad news, he wasn't sure yet which: He jumped in and gave Big Bill Haywood some advice. The man liked it, enough to personally ask for more help. And for loyalty to the union.

Overpowered and feeling at swept along, Ben nodded.

"Sure, Bill. You can count on me."

Ben didn't stop to think. The plan had been to put off going back to the mines for a few days. He needed to tie up loose ends from the Emma May, make arrangements for living space, and catch his breath. He was between jobs, so to speak, and needed to get settled, make arrangements, and so forth. Not enjoyable stuff, but necessary. And God knew, he had put heart and soul and every waking hour into that mine, and he needed a little down time, rest.

"Bill, I am washed out and torn up. I'm just plain tired. Before I go back, I need a few days to get collected."

Haywood would have none of it. The man insisted that Ben had to go get working in the mines right away. The big man acted almost as if the WFM would collapse if Ben took a few days to get started.

As usual, what Haywood wanted, Haywood got.

"I need you to get out there, Ben. I need you to listen, and pay attention to what the foremen and superintendents are doing. And what the men, the miners, are feeling, thinking, talking about. We have no time to lose." He looked at Ben, almost daring him to say something.

"Good. I'll have Shuler call in a favor. You'll have a job tomorrow. See him first thing."

VII

PER LAST NIGHT'S INSTRUCTIONS, BEN ROSE AND WENT TO SEE Shuler who finagled a job for him at the Findley Mine.

He found himself crowded into the downward lift, ready or not. The bunch in the cage were unfamiliar to him and vice versa. No one talked to him, nor even gave him a second glance. He felt extra bleary and on edge. His night's sleep was shorter than a tomcat's yowl. And he was the newbie. He didn't bother to mention his mining experience—the crew didn't care. He'd have to prove himself. Again.

Of course as the new guy he got the scut jobs. That, he expected. He had forgotten that no one would talk to the newbie more than absolutely necessary. He was philosophical about it, nothing new or different with that. That is how it worked with a crew who depended on each other for their lives and safety. He would have to put some time in. It would take a week or so to show his competence and at least be accepted. It would take longer to be trusted.

Ben felt washed along, out of control. He was happy to work and take the pulse of the workers and the mine managers. He just hoped that what he reported didn't encourage anyone in the union to do anything rash. From what he had heard of Big Bill, he wasn't entirely sure about that.

THE CAGE THUMPED TO A STOP. BEFORE THE DOOR OPENED, HE thought not of the coming day, but of Abby. She said she was a teacher. There weren't that many schools in town. In the next few days he intended to look her up and get acquainted. Sure, Haywood wanted information, and he'd do what he could for that. But his all day every day wouldn't be given to the WFM. After all, he had a life and intended to enjoy it, be sociable and enjoy what the town had to offer. This woman he just met intrigued him. A smart independent woman who was interested in mines and miners was a find in any case. Her being drop dead gorgeous with brown eyes and red hair was icing on the cake. He looked forward to getting to know her.

The men filed out of the cage. He figured he probably had more experience than any two of them combined. Still, no point in making enemies. He went out of his way not to make waves. As the newbie he waited his turn and was last to exit.

"Come on out, Bob." The crew chief stood out and gestured for him to get with it.

"Oh, sorry. I was waiting for the car to clear and got to thinking of something I need to do tonight. And my name is Ben."

"Alright, man. Ben it is. No more wool gathering, Ben. Let's get to it. I need you to…"

UP TOP, LATER THAT DAY. ABBY THOUGHT BACK ON THE UNION meeting. Haywood sure was pressing hard to cut the members out of voting on decisions. That sounded good in theory. Sure, the local elected officials could act quickly on behalf of their members. But maybe the members didn't need quick action. If something severe or awful happened they could vote easily enough. She had qualms about it all. But Haywood pushed them through and a new regime was running the WFM. Members were needed only to pay dues and man the picket lines. Apparently their opinions were not wanted or needed. To organize her thoughts and express herself, she liked to talk things through. Lon had the pleasure of hearing her thoughts on the WFM. Not that he cared for the subject, but as an owner he found it smart to know what they were thinking.

Abby came home from work and was running her thoughts and feelings and impressions. talking to her brother.

"So, Lon, it looks like this has gone through, and is the new way of business for the union. I wonder. The question is, who will direct the local officials if they aren't listening to and getting consent from their members? Will they get marching orders from Denver? Do you think they will talk to other locals, or what? Wasn't that the point of having a local chapter with its own officials, to keep control local? I guess that is a thing of the past for this union, at least for now."

"I expect, little sister, that your Big Bill Haywood up north will be pulling the strings. Him and the other muckety-mucks at the Western Federation of Miners. Sitting up at national headquarters in Denver, not here in Cripple Creek. They'll be making the decisions, I'd bet the house. Why else would Haywood push for the change? He wants control. I'm afraid he wants to push for trouble, you mark my words."

"Do you think? Trouble? Would they override the local members? What kind of trouble?"

Lon frowned. "In a heartbeat they would. Haywood and his cronies would love a strike, a big messy bloody headline grabbing strike. And they—he—won't much care who down here gets bruised or shoved around since they will be looking at the big picture, so to speak. He'll do his best to make it look like the poor miners are being squashed by the big bad owners."

"That doesn't sound good for us here."

"Maybe not. But why do you care, Abby? You are paid by the school, come what may. All you are supposed to do is teach arithmetic, reading, geography and grammar. Right? How does Big Bill Haywood taking control affect your classroom and the abc's?"

"Well, many, perhaps most, of my students have a father or brother in the mines. I care and they care because the men in the mines—and people everywhere else—should be treated right. If we treat our workers right, everyone wins. They are not chattel, to be used and discarded. Yes, I teach children, and I do it well. But there is more to life than helping a child add a column of figures or learn to spell."

"Fine sentiments. Being treated right doesn't mean letting yourself be pushed around by Big Bill and his cohorts in Denver. Has he asked the miners if they are happy with their jobs?"

"I don't know about that. Maybe. The thing is, if there is a strike or problems, my students will suffer. As will the rest of the town. That, I don't want."

"Nor do I. These are hard men, Abby. Your guy Haywood has men at his call for dirty work. I don't mean dirty like grime, I mean like hurting people or worse. He has used them in the past and will again. And the mine owners will return like for like. They too have men in the wings who can break bones. Hell...Oops, sorry, Ab. My goodness, I should say, not the other word. Anyway, it almost seems like there are private armies out there just itching to have at each other."

He smiled and put an arm around her shoulder. "Just please be careful with this, this union involvement of yours. These men play for keeps and I don't want you to get in the middle of things."

"Things? What things?"

"Problems are building, Abby. I talk to miners and merchants as well as bankers and mine owners. On any one day, I mix with all kinds of folks. Seems to me I have a pretty good feel for what is really happening."

He smiled, glad to tell a fun story rather than discuss union issues.

"The other day I was out knocking rocks and got to talking to an old timer. Guy just kind of turned up over on the hill. We chatted, light stuff, you know, neither of us really letting on what he was doing up there. From what he said, it was clear he knew mining and the area. Turns out it was Stratton! The mining millionaire, out looking for color on a Cripple Creek hillside! As if he needs to find more good ore! He told me he just loves walking the country. After we chewed on mining and metals and so forth, talk turned to local issues. He said the WFM and the MOA seemed on a collision course. Said the damn fools—sorry, Ab—said that neither side much cared to talk. Both are willing to put up the dukes and each is sure they can prevail."

She was nonplussed. "Winfield Scott Stratton said that? You really talked with him?"

"Yes, Abby. Winfield Scott Stratton. Now, think about it: the man has connections and knowledge you and I can only dream of. If anyone knows what is happening around here, it is he. He sees storm clouds abuilding. So that is why I want you to be careful."

"I am and I will be careful, Lon." She sounded grave and a little shaken, and in fact she did feel that way.

Lon had a tale to finish. "By the way, he told me a story about his name. I just love why his folks gave him that first and middle name. He was named after a successful US Army general. The man who led our forces in the Mexican War was Winfield Scott. The man who needed a hoist to get on his horse by the end of his career, he was so stout. I think fat is the proper term for that. Anyway, I guess Stratton's parents hoped the name would inspire him."

"Well, he sure is a successful miner, so his parents' wish came to pass."

He shook his head and smiled, almost as if glad to be done, and he referred back to the local scene. "True. But I digress. My point is, things are not good. The MOA and WFM seem to be on collision course."

"Speak English, Lon. I ask again, what things?"

"Things—I mean the overall situation here. I hope I'm wrong, Ab. But I think in the next year or so we will see fights, lockouts, strikes, bloodshed. Killings and beatings. Missing dynamite, maybe used to harm and intimidate. The time for talking will be over, I'm afraid. More than half the cause, the problem, is these union people you go to meetings with. Be careful, and be afraid, my sister."

The first day's shift always felt long. Today was no exception. Coming out of the shaft house, Ben was glad for fresh air. He stopped to take a lungful in and saw his old friend Juni. The big man had a shiner and favored one arm, holding it close.

Seeing the big guy stove up like that for some reason amused him. He laughed. "Good Lord, Juni! Did you run into a windmill or something? Looks like you got the worst of it. Seriously, what happened? You fall down a shaft or maybe you went down the wrong alley?"

"Hi Ben." He looked around, making sure no one could hear him. He adjusted his arm, grimacing as he moved it. "Oh Ben. Do I have stories. It has been a while since we talked, no?"

"Yeah. What have you been up to? What mine are you at?"

His friend put on a smug grin. "Not at a mine at all. I've stopped working as a miner. What happened is, well, you know, I was talking about making a change. So, I am a high grader. I told you about that, didn't I? We, me and my friends, we have a few abandoned mine entrances no one else knows about. And we use them from time to time after hours."

"We? Oh yeah, that Henry Orwell or something like that, the guy you were telling me about. So why the bruises and tender arm? Did an old entrance cave in on you or something?"

"Harry Orchard is his name. Good guy. But don't tell that to anyone. He likes people to think he is a first class bastard. Anyway, what happened, Ben, is complicated."

"I'm all ears, Juni. Complicated is nothing new around here." As he said this he thought about wanting to get to know a woman he met, and he wanted to tell Haywood the right things, and he wanted to get the Emma May mess straightened out, and now he was worried about Juni on top of everything. Nothing easy about all that!

Juni nodded.

"Yah. Well, we—Harry, me, and some others—have good access to good ore. You know about that. And we have some assayers, over twenty of them, who will buy our ore. At a discount, but that's how it works."

"How do you know you're getting what it is worth? How do you trust those guys, or can you? Some of that ore you take out on the sly has to be hot, real valuable stuff. It must have thousands of dollars of gold per ton, so a few hundred pounds would be a good day's effort."

"Exactly, Ben. That is where the hammer meets the anvil. Most of the time things go along smooth, no problems or funny business. We give them ore and they pay us a fair but discounted amount. But, the thing is, we found out that one of our assayers was shorting us. No doubt about it. Now, they pay as little as they can get us to take, that we understand. And we understand that straight up mistakes are sometimes made. Well and good."

Juni adjusted his arm again. "Damn, I'll be glad when this heals up. Anyway, we found this guy was using funny numbers, and was flagrant about it. He was shorting us by half or more, time after time. And worse, he was starting to talk about it, bragged on the street. Word was getting out that we wouldn't react if he cheated us. We would just let it go, he thought."

"So what happened, my big easy going friend? Were you suddenly not so mild mannered?"

He grunted. "Kind of. What happened is, Harry had me pay this guy a visit. Said I was big and intimidating and should go teach him a lesson. Which I did. Thing is, he got one swing at me with a billy club before I got him down. Like you said, someone got the worst end of an encounter."

The big man stood up a little taller, his grin turning wicked. "I'm here to tell you it wasn't me. And I can guarantee that all the assayers are talking now, yes they are. And I am sure they will be extra honest with their weights and analyst reports from here on in."

"Man oh man, Juni. You are getting in deep."

"Well, Ben, it is a good living. I don't have to worry much about a roof cave in and not at all about a premature blast. I don't have to put up with the mine owners or a foreman telling me what to do all day. Matter of fact, I'm sticking it to them! I'll take this any time."

"Wow."

Juni was on a toot. "Plus Harry is introducing me around. He has some stories about his time up in Idaho—he got into some big time fights. He's a good man to have in your corner. I'll introduce you two if you want to get out of the mining rat race."

"No, I don't think so, not just now."

Ben flashed on Big Bill Haywood saying he wanted to know what was happening out in the camp. This was definitely part of the goings on in Cripple Creek and Victor—might be worth learning more about.

"Well, on second thought. Maybe sometime, Juni. I doubt I'll want to get into high grading or such, who knows. But I would like to meet your friend."

"I will be in touch. We're busy now and I need to be careful about bringing people in, if you know what I mean. So it won't be for a while yet."

Ben nodded.

"Alright. Take care of yourself, big man, and we'll see you around."

Juni nodded, smiled, went down the road.

Ben stayed there, marveling at how Juni seemed to love his new murky world. Then he decided what the heck, the guy seemed happy. He thought on his day at the Findley. He was still in work clothes, tired and more than a bit dirty. But he wanted to go see her, the redhead from the meeting. No time to get cleaned up now, not after his stopping to talk with Juni. He would go try to see the woman he sat next to at the meeting, Abigail. Abigail Bosini, she said. Call me Abby, she said. He would, he would.

She didn't say where she lived, of course. But she did say she taught at the school. And clearly she was open to getting acquainted. Best he could do is swing by the school. No doubt the students were gone for the day but with luck, she would still be around. Ben took a deep breath and headed off.

She was still at the school. There she stood, in front of the classroom building. Her red hair billowed and glowed. It brought to mind a distant forest fire after dark he had once seen. But she was talking to some man in a suit. And it wasn't formal conversation. Ben could tell they knew each other well by the way they stood and looked at each other. Not good.

Disappointed and unsure, he decided to trudge on. If he turned back he'd draw attention. He planned to go by as if he hadn't noticed them. Be just what he was, a scruffy miner on the way home after the shift. He kept eyes on the street in front and glumly traipsed on.

"Ben! Ben McNall!"

He took another step, amazed that she would remember him. Then stopped, turned.

Ben walked to the pair. He couldn't bring himself to act surprised. "Well, hello, Abigail."

The man gave him the eye, no doubt wondering who is this guy? How does he know my woman? Ben returned the gaze, confident but not challenging.

She took half a step away from the man towards him, smiling.

"Ben, meet my brother Dillon Bosini. Lon, meet my friend Ben McNall."

Both men worked not to change expression at the term 'friend.' Neither reacted but both were surprised, almost astonished.

Lon wondered where on earth she had made friends with a miner. Oh yeah, he realized, probably at one of those damned Union meetings.

Friend! Ben was on cloud nine—he was head over heels for this woman when they met, and she considered him a friend! It was all he could do not to smile and giggle. He stifled it, put on his friendly smile, and extended his hand.

"Nice to meet you, Mr. Bosini."

"Mr. McNall." Lon nodded as they shook.

Abby saw two men shake hands with the requisite smile. But the image she got was of two boxers tapping fists before going after each other.

She spoke to cool the hostility, wanting to let them get acquainted. "We were just going for a cup of coffee at the Continental Hotel, Ben. Would you like to join us?"

Lon glared at her. Ben almost grinned, images rushing at him. He pictured himself at the Continental, Cripple's finest hotel. He would be grimy, sipping coffee out of fine china, people staring, the waiters and Lon giving him dirty looks, Abby smiling and making small talk. It would give the gossip mill material, for sure!

"Oh. Yes, thank you, Abigail, I'd like to do that. Sometime." He grinned, looked down at his dusty clothes, and deadpanned, "Today I doubt they'd welcome me with open arms."

"Call me Abby," she interrupted.

"Abby, sure Abby. Thanks, really, but I need to get home. I'm still in my work clothes."

Lon snorted at that; she glared at her brother. Ben ignored the exchange.

"I like to be cleaned up before I go out. Another time?"

"How about tomorrow evening?" Lon was surprised at her assertive response. Somehow Ben was not.

"Sure. I'll come by here about the same time tomorrow. How does that work, Abby?"

"I look forward to it!"

Ben smiled as he looked to Lon and stuck out his hand. "Mr. Bosini, nice to have met you."

"Oh call him Lon." Abby smiled at her brother. "Mr. Bosini sounds pompous. He's not stuffy, are you, Lon?"

Lon was not amused, but smiled anyway. He took the proffered hand. Trying not to sound stiff and 'pompous,' he responded. "Ben, isn't it? Nice to meet you too, Ben."

Brother and sister watched Ben lope off, a spring in his step. They spoke first thoughts at the same time.

"My God, Abby, who are you getting mixed up with? I hoped he would stop by. I wanted to talk with him."

They looked barbs at each other, then relaxed and laughed.

"I'm not a baby, Lon. And I know what you're thinking, but you need to mind your own business."

He held his hands up in front, in surrender. Time to go with the flow.

"You are right, Abby. I was surprised is all. I have to say, he seems a reasonable fellow. At least he speaks English and can put together a sentence. Unlike some miners. I have to say, he seems familiar somehow. I feel like I have met him somewhere."

"Well, Lon, he is a miner. And it is no surprise that he dresses like one. You may well have seen him on the street or in a store or somewhere. Actually, I hear that he has part interest in a claim, a working mine. Come to think of it, he told me the name is the Emma May. He's into several mines somehow, I guess."

"The Emma May..." Lon rubbed his chin.

"Oh yeah! He's the guy with the shotgun! That's it, he was the guy I crossed swords with when I accidentally came onto his turf. Up on Strombo Point. That was the name on the claim, the Emma May, I remember now. That was not long before I found color and filed my own claim. In the same area."

The puzzle pieces started to fall into place for Lon. He asked, "So his operation is going well?"

"Yes. No. I don't know. I'm not sure how it is going for him. He said something about being between mines or some such. And he was sure working in a mine somewhere today. No doubt I'll learn more tomorrow."

Her smile disappeared. "Why? Will there be conflict with our Double I?"

Lon looked to the horizon, thinking quickly. He turned back to Abby.

"This stays between you and me, sis, alright? Just you and me. Promise? At least until we sort things out, don't tell him or anyone."

His expression caught her attention; he went on.

"The thing is, yeah, there could be conflict. It is possible, likely even, that his Emma May and our Double I are looking to tap the same vein of ore. If so, there could be problems. Even lawsuits. We need to find out. I'll have people discreetly look into it. Plus, and I hate to say this, but you need to know. Remember I told you some crazed angry miner leveled his shotgun at me, running me away?

She nodded. "It was him. He ran me off what he called his claim with a shotgun, a loaded shotgun. So...Please be very careful what you say around him."

He looked her in the eye, making sure she understood. She nodded doubtfully and he went on.

"At least let me check on some things first. I'll see what I can find out about him and his mine. And maybe I can stop by tomorrow at this time or a little earlier to fill you in. Alright? I was going to nose around Victor tomorrow. But that can wait. This is more important. No time for coffee at the Continental, sorry. See you tomorrow, gotta go."

His posture and pace made her think of a grizzly following the scent of an elk.

The next afternoon, Lon came to the school. On the front steps, the view was fine. Cripple Creek was in a big bowl, an old volcanic caldera or crater he had learned. There were buildings all around, mines, smoking chimneys from stoves and kilns and smelters. It was a busy, dynamic, exciting place to be.

He walked across the road and inspected the school building. It was sturdy, three stories of brick, a fitting monument to the importance of education. He was glad there were children in the camp, and he was proud that his sister was teaching them. Somewhere a bell rang and kids started running out onto the street, heading home or to work or to play. Soon Abby stepped out.

She nodded at Lon, hoping he would be quick. She really preferred that he and Ben not meet again today.

"So you did some checking?"

"Yes, I did Abby. Good news and bad news. The good: Ben is on the up and up. He really does own an interest, one third, of the Emma May. It started fast I'm told, ore with valuable content. Typical story, one of hundreds. He was, is, in with another owner, a man named Eisner, Cook Eisner. They put everything they could into it. Like most every other hole in this County, lots of money in but precious little good ore out."

"That doesn't surprise me. I had a feeling about him..."

Lon talked over her. "Now, the bad: word is their vein pinched out. The claim is worthless or damn near so. Pardon my expression. I guess both Ben and this Cook fellow are washed up. At least as of now. Ben borrowed on his house and has lost it. Which is why he was dressed as a miner. I'm told he has gone back to wage work."

She nodded. "Oh. That is not good, but like you say that story has been told over and over in this district. Many men come back and make the best of things. I imagine Ben will too. Anyway, what does that mean for us, the Double I?"

"First, I hope our vein doesn't pinch out or disappear like theirs did. Knowing theirs did lets us know we need to check some things we might not have. But if it does? Well, like you say, it has happened before and will again. If not, well, that's life in the gold camp. Their misfortune is our good luck. Probably then we will suffer no disputes, at least from them. No lawsuits or loaded shotguns! Still, be careful what you tell him. I'd ask you not mention the Double I, at least not until you get to know him a little better."

He looked at his watch. "Well, I'd better run. I don't want to interfere with your social life!"

He smiled and she knew he wasn't being sarcastic. Hard as it was for him, he really meant it.

"I'll see you later, Abby. You're the only sister I have here, so take care."

As he walked away he pondered. Could he buy the Emma May? Or somehow tie it up? Maybe he could lowball a cash offer. That would help him and Abby. It would help Ben which Abby would like. And best of all it would put a stopper in disputed geology, claims, and mineral rights, at least from Ben and the Emma May. He'd have to give that one some thought, and quick before someone else did it.

The day before, as Ben walked home, something gnawed at him. That guy, Lon, was someone he had seen or run into. He couldn't put his finger on when and where. Still, he was sure they had crossed paths. He just hoped it wasn't a hostile encounter, rather just saying hello in the street.

No matter. Today he looked forward to seeing Abby. His being a brother explained the way they stood and looked at each other. He could tell they weren't strangers when he first saw them, and was glad they were siblings. Abby clearly didn't take orders from him no matter who he was. In his mind's eye he saw her hair and brown eyes which almost caused him to miss the turn to his place.

Not concentrating on it let his mind come forth with the story on Lon. He was the man nosing around up on the claim, one of many. A while back. Ben ran him off with the shotgun. As he thought back, Lon didn't act ornery or try to make trouble. Apparently the guy really had just been walking the area and inadvertently crossed the line. Ben wondered what he was up to, what work he did. Since he wore a coat and tie he wasn't a miner. Abby would know.

The second day with a crew usually went faster than the first. But the shift seemed to last forever. Ben figured it was anticipating seeing Abby, not the work itself. The cage finally lurched to a stop on the surface. The door opened to fresh air, always a pleasure. He intended to rush home, clean up, and change.

Maybe not—there stood Shuler. He fell in next to Ben and they walked.

"Ah, Mcnall, there you be. We need to talk."

"Can it wait for a day? I have an appointment."

"An appointment?" Ben could practically hear the man's eyebrows raise as he blurted his reaction, part question and part snark.

"Well, not really an appointment. I'm to meet a lady friend."

"Oh, a lady friend. I see. Well, your 'lady' can entertain her next 'guest' and you can go 'meet' her later."

Ben stopped, faced him, anger building. "She's not that kind of a lady. I'm happy to talk with you but not tonight. Tomorrow, here? Will that work?"

The Union boss was almost amused at a miner getting mad and talking back. It was refreshing in a way.

"Yeah, sure. Go enjoy your 'lady friend.' Come by the union offices tomorrow after shift." He smiled sarcastically and strode off.

Shuler delayed him just enough that he didn't have time to go home and change. So he dropped in on a friendly barkeep who let him use the back room to splash water, clean his face and smooth his hair. His shirt and trousers he dusted off best he could. It'd have to do.

Abby stood in her empty classroom, looking out the window. The kids were gone and the building quiet. She stayed back, out of sight of the few people walking by. There! He slowly, almost shyly walked up to the main door. She opened it and stepped out.

"Hello, Ben. Nice to see you again."

"Good to see you too, Abby."

A brief awkward moment, broken by his suggestion. "Let's walk. The day is pleasant. And I am only halfway presentable. Shuler waylaid me and I didn't have time to go change."

"That's alright. You look fine. Yes, let's walk. Fresh air feels good after a day shut up in the classroom."

He smiled. "So true. I bet your classroom air is nicer than my air down in the mine. Not to complain, just saying!"

She shrugged. "You haven't smelled a roomful of sweaty boys and girls, giggling and trying to get the attention of someone across the room! It can get gamy. Maybe not to the level of a mine, but kind of shall we say aromatic."

"I never thought of that."

"Nor had I until I faced it the first time. So how was your day in the mine?"

"Oh, fine. Usual day, especially since I am the new guy on the crew. Can't complain. Say, I hope I didn't offend yesterday, being in work clothes and all. Lon seemed a little standoffish."

"Just surprised is all. And he doesn't own me and what he thinks isn't important. Why do you care?"

"Hey, we're off to a bad start. I'm just trying to make conversation. Maybe I'm not the best at that. What I care about is what you think, not about anyone else. Peace?"

Laughter. "Sure. I need to take things less seriously, really I do."

He couldn't resist. With a grin, "I can see you are serious about that."

A smile. "You got that right buster, and don't you forget it! Seriously...." At that she paused, made a face and laughed; he did too.

"What I mean, Ben, is let's get past these silly word games. Tell me about yourself. Where did you come from? How do you like Cripple? You mentioned something about two mines or between mines? Did I get that right? I think you said something about that before Big Bill started his tirade at the meeting."

"Yes, two mines. Long story. I am a hard rock miner, Abby. That is what I do, who I am. Quite a change for an Ohioan, or at least I was born there. Left as a lad. Silverton and Telluride and Aspen and Creede and Breckenridge and you name it, I mined or looked for mine work at all of those and more before I landed here. Have worked mines since I landed in Colorado after a hobo ride from Ohio. I saw enough of trains and tracks and railyards and jail to last me. I am around mines whenever I can. Love it. And now here in Cripple I own one third of a claim, the Emma May, over on Strombo Point. Me and Cooker Eisner."

"Can't say I know the man. But that is exciting. Tell me about it. Who is Emma May and is she producing good ore for you?"

"It certainly started out well. Cook found good ore, filed a claim. I staked him to start development. He and I put everything, heart, soul, time, money, everything into it. The vein assayed high and we were getting after it. Life was starting to look pretty cushy there for a few weeks."

He paused. "Now, it seems to be ending. Correction: it has ended. And not well. From all indications, the vein—the ore bearing rock—has pinched out. We've looked everywhere across the claim, every way including sideways to find it again, but no luck. The Emma May Mine is, as the Germans say, kaput."

He scuffed his feet and fought a black fog falling on him.

"Oh, you ask who is Emma May? She is Cook's girl back east, I think in Indiana. She won't be coming west any time soon, I'm afraid. Too bad. She was starting to get excited about it, I guess."

"Oh? The old vanishing seam problem. Too bad. I assume you two have driven drifts and stopes and run analyses on every rock you can find. Still no luck finding where the vein is?"

"My, for a children's teacher you know quite a bit about mining, don't you?"

"I'll thank you not to condescend." She paused, a chill in her expression.

A warm smile, all forgiven and point made.

"This is a gold camp, Ben. Mines and their components are topic number one everywhere. And my brother talks geology and mines like there's no tomorrow. It is what we hear from most everyone we know, and probably everyone you know. Even my schoolchildren talk the lingo."

"Point taken. Like I say, I'm not the best at conversation."

"No problem, Ben. So, no headway in relocating the vein in your claim?"

"No."

He didn't try to hide the disappointment and bitterness that the Emma May left in her wake.

"The vein pinched out. It is not to be found on our claim. We have ceased work, let our small crew go. Sad. Hey, do you know anybody who wants to buy a worthless claim? I know of one with a fine name and a shaft house, a shaft to nowhere."

"Ah..."

"So, the upshot is, I am back mucking ore on the three dollar a day miner's wage. I'm just another joe, a hard rock miner in a gold camp."

"Oh. Not to be harsh, Ben, but this is not a new story. That doesn't take the sting away or make it easy, I understand. Same thing has happened hundreds of times here in Cripple, thousands of times in Colorado alone. Some men throw in the towel. Some dust themselves off, go out, file another claim, and make a go of it."

"I suppose. But that's a tiresome and sore subject. Enough about me and mining and the Emma May." He stopped, turned, and let her know she had his attention.

"Tell me about Abby. How did she end up a school teacher in a remote gold mining camp? What does she like to do when she isn't teaching or going to union meetings? That is pretty unusual, going to union meetings, isn't it? Does the school board or superintendent say anything about that? What do your students say about the owners and the union?"

"What I do on my time is my business. Anything done after hours doesn't seep into the lessons."

He barely heard that.

"Ah, I see. And what does she have to say about David Moffat trying to build a railroad direct from Denver to Salt Lake City? Or the Congress sending our armies to invade Cuba and the Philippines? How about those Wright brothers with their flying machine? Can you imagine soaring like a bird?"

Smiling, she held up her hands. "Whoa, Ben. I can't drink from a fire hose! One thing at a time!"

He simply smiled, and made a 'come on' motion, waggling his index finger as if to draw her out.

"You want me to tell you about Abby. Well, I teach because I love learning and love helping people, children, to learn. Lon and I grew up in a home where books and learning were valued. My parents made sure we both got an education. We were encouraged to read and express ourselves. I think those skills are important and I try to pass them on."

"Well your parents sure did well with that!"

She smiled at that. "Yes I think they did. You know current affairs, it seems. Here's what I think: Why do we need to invade Cuba? Or the Philippines? We Americans don't need them do we? Are they worth Yankee blood? I have trouble with that, but no one asked me. Since women can't vote in national elections it doesn't seem to matter. At least we have a voice here in Colorado elections." She scuffed along silently for a few steps.

"As to the Wright's aeroplane—wonderful invention, what ingenuity and talent! I wonder how their mother and sister influenced them. The 'plane will change our world. It will alter everyones' lives in ways we can't even imagine. I hope I get to see and ride in one someday!"

Frankly Ben wasn't prepared for such a thoughtful and informed response. "Am I walking with a high school graduate? Gee, I barely made it through the sixth grade. Maybe I should walk three paces behind..."

"Oh stop, Ben. You are clearly smart and educated. Schooling is important but what one does with one's mind after is even more so."

"Well, maybe. Abby, I am curious. Well, first, it never occurred to me that women would want to vote. But thinking even a little bit about it, I can see why. And, I'm a little surprised you aren't more angry about it."

"Oh I am not happy about it, and hope I live to cast a vote for President. What is second on your list?"

"I'm wondering about your interest in the union and labor..."

"Look, most people spend more hours at the job and with coworkers than they do with family. At least more waking hours. Right? So why is it so strange for me to care about how people spend their work days? I think people ought to have a reasonably safe place to work and be paid a living wage to do it. That's all."

"That aspect never occurred to me, Abby. Very creative."

"Condescending, Ben, don't. But, let me back up. I guess I'll take that as a compliment. What I care about, in Big Bill's lingo, is that workers need to resist becoming just a cog in the capitalist machine."

He laughed. "That aspect has occurred to me. I can't disagree."

She stopped, looked at him. "Ben, I am interested in union meetings because they are important. Workers need bosses and bosses need workers. For the life of me I don't understand why they act like enemies. They are working to the same end, right? A profitable and stable organization, right? It makes no sense, none at all. I simply don't understand. They should be, could be, allies if they would just listen and talk to each other."

They started to walk again. "Ah, the eternal question, Abby. Better men than me have tackled that one. Maybe we can discuss it some other time. How did you end up in Cripple?

"Like I say, I got an education. Grew up in Kansas City, which was nice enough, prosperous and proper and so forth. My family has produced river pilots for generations. Being a pilot's wife, always waiting for him to come back, was not for me. I had to get out. Too flat, too humid, too many trees to see anything. Life is too short to have your only concern a husband and the children."

"No danger of that, not with your opinions and outlook."

She ignored that. "I had a decent education, thanks to my parents. I heard of the need for teachers in Colorado. My brother was taking some class work at the Colorado School of Mines and I wanted to be around him if I could. Plus, Colorado seemed a glamorous and exciting place. I decided, 'why not?' So I came out to Golden, stayed with Lon, and scared up jobs to apply for. There were lots in the area but Denver is just another city. I looked further afield. Especially to the mining camps. Not sure, but I think I was the only applicant for the job here."

"You were clearly the most qualified, else they wouldn't have hired you. Say, not to change the subject, because I love to hear about you. And I want to hear more, yes I do. I forgot, and I really meant to tell you right away. You mentioned your brother Lon. I think I have met him. Before yesterday, that is."

"Oh? Actually I think he mentioned that too. Something about you running him off your claim, with a shotgun." She smiled, joking. "Now, that wasn't a neighborly thing to do, not friendly at all. Is that any way to greet someone here in Cripple Creek?"

"Seriously—there's that word again. I mean, Abby, you would be surprised. There are many skunks lurking around just waiting for a chance to take your claim. Cheat or steal or fast talk or shoot, any way they could. You can't be too careful, you really can't."

She arched her eyebrows, drawing his eye to the hair again. His mind seized up until she cleared her throat.

"I mean, put yourself in my shoes. There I was. I had what seemed to be a good prosperous mine. And up the hill here comes some joker, ambling over my property, my claim, chipping at our rocks! Heck, some would have shot him on sight for that! I didn't know what he was up to. Turns out he was just looking around. But I have had men try to pull a gun or swing a club in the same situation, even nearly the same place. That day Lon clearly wanted no trouble and he left cheerfully and willingly. I was damn glad of it."

She smiled. "Good thing you didn't shoot him—that would have gotten us off to a bad start."

"Ha. I mean it Abby. Mining is a serious business. A few days before your brother, I had to shoot a barrel of buckshot over the head of some intruder. That got the man's attention and he ate dirt real quick. After I let him get up off the ground he left in a hurry. Not cheerfully but fast as a mama bear protecting her cub."

"Oh Ben. What if he wouldn't leave?"

"Don't know. I'm not sure I could have pulled the trigger, emptied the other barrel at him. Thank God he wouldn't take that chance. My point is, if you have a claim you have to protect it. However you have to, every way you can."

"Well, I'm glad you didn't shoot him. Lon, I mean. And I know he wants to talk more with you."

"He doesn't approve of me does he." This was not a question.

"Well, he is wary of the union. And he is a protective older brother, not that I need protecting. He isn't so much wary of men looking after themselves and their jobs, really he isn't. He is wary of the union itself. Some of their actions are violent and mean."

"I understand that. But, just like a prospector has to protect his claim, so does a worker. A three dollar a day working miner has to protect the job and working conditions and his safety. Just as diligently as the prospector. For that matter, I can see why the owners are so jealous of their hiring and their profits. This is a hard, competitive business. Everyone is looking at every angle to get along efficiently and make money."

Ben's chuckle was unamused. "Like I say, it is a hard business. I know more about that now than I ever wanted to."

"So it is. Perhaps we three—you, Lon and I—could break bread this weekend. Sunday?"

"Thanks, Abby. That sounds good to me. I'd like to do that. When and where?"

THE THREE GOT INTO A RHYTHM OF MONTHLY MEALS. THE gatherings let them talk, listen, provide and receive support. There were disagreements and sometimes heated words, but it was in a sense family time.

Over a year later they sat for one of their regular dinners. It had been a little while—the previous month simply hadn't worked out. Abby looked forward to it. As usual there were arguments and differing points of view, but usually they were smiled away, agreeing to disagree. Occasionally someone, usually her, had to referee some seemingly serious difference. Often as not, a week or two later the disagreeing parties couldn't remember what they were steamed about. She wondered if there would be sparks this time.

It was a midwinter's evening. The early dark, snow on the ground and wind gusts made gathering with family extra comforting. A good meal and warm room with people who care make up for much. Abby was describing some of her classes and what she was teaching.

"My social studies class learned about the formation of our new county, Teller County. That was back in March of '99. Interesting lessons in politics, warts and all. Teller was carved out of El Paso County, the big county surrounding Colorado Springs. This happened to the consternation of the millionaires down there. Up in Denver, the Legislature passed an enabling law despite their long and noisy opposition."

"Once the law allowed it to move forward, we here had to form the government. The students observed and studied how local factions pulled for Victor and Cripple Creek to be the County seat. It was a good practical lesson in civics and government. They saw how the courts, sheriff, and all the other County offices were filled by locals, not people beholden to Colorado Springs interests. They were Cripple Creek and Victor men, mostly miners. Which is what the locals here wanted."

"I'm sure you made it interesting and meaningful, little sister."

"It is meaningful, Lon. Knowledge of how our system works makes for good, informed citizens and voters. If the county commissioner is a neighbor, it is less likely he will push through rules the citizens don't like. And don't call me little sister." Ben smiled at her feisty rejoinder. Lon just smiled and shook his head.

She continued.

"And it is important because it brought our law enforcement and government close to home. Like I said, it used to be the money boys in Colorado Springs running the show. Not so much now in Teller County. Now those holding the reins are miners and merchants and local people and they all understand our situation. And some of them are millionaires, yes, but most of those were plumbers or barkeeps or carpenters first, and they still remember life as a worker. Plus, like I say, we are their neighbors and they can't hide away in a big city. They have to face us every day."

Lon's smile went away. "Ah, the working man ploy. You use it often, don't you Abby. I am all for the working man, the guy who actually gets his hands dirty or keeps the accounts or runs the tests or sells the hardware. We can't do without them. They are the backbone of our communities. And they have a history, do such Americans, of planning and acting for themselves, individually. They don't take kindly to being told what to do. Be it from a tycoon, a politician, or a union boss, they don't like it."

"Historically our system rewards and fosters people thinking for themselves. A union by definition says people are too weak and maybe too stupid to do that. And I think that is a problem. Plus, some unions have a bad attitude. They, or most unionists, seem to think that anyone who owns a mine or a bank or a store is a greedy filthy bloodsucking capitalist who will step on a worker for even a dime. That is untrue and is on the wrong track entirely. It is not helpful for such lies to be spread. I just hope you don't repeat those stories to your students."

"I do nothing of the..."

Ben interrupted her. "Lon, I agree the Union hardliners sometimes start to believe their statements. Some even have weird or lopsided ideas about the economy and workmen. But I have to tell you, the typical miner does not think a businessman is bad. He knows better. The mine man is most interested in a steady job and taking care of his family. No more no less. The union guys from Denver calling businessmen villains doesn't get the job done, and frankly isn't helpful."

"Well, Ben, you're in a position to know what the guys out there are thinking. Since you are still working an eight hour shift for the standard $3. I don't understand why you do that."

Ben shrugged. Lon went on.

"I mean, you and I could have worked a deal: I take over the Emma May, fold it into the Double I."

Abby jumped in. "You mean, you and I could have worked a deal with Ben. We take over the Emma May, not you."

"You're right, Abby. Figure of speech. I should have said we."

He turned back to Ben.

"Had we done that, you and Cooker would have been able to clear your debts. He'd be out of the picture but you would have a one sixth interest in the Double I. But you passed on that, took your lumps, and went back to being a wage slave. Why? I still don't get it."

Lon glanced at Abby as he said this. The two of them had talked and even argued long over this. For their own reasons both wanted Ben in on the deal. Abby particularly wanted to have Ben share in the ownership. After Lon and

Abby agreed what to offer, Ben simply shook his head no. He had different ideas and wouldn't agree to it.

"So why do you still do wage work, Ben? I don't think I have ever asked you directly. Does your foreman or your friend Big Bill Haywood think better of you for not being a nasty bloodsucker?" Lon said this with a smile in his voice.

"You ask why, Lon? Because I don't want to hang around with the men who are owners. Present company excepted, of course! Seriously, I am simply not interested in being a mine owner. I have been in those shoes, done it, and don't want to do it again. To put it in a few words, I am a miner."

He shrugged. "And, really, Lon, I mean no offense. You're not like many of them, owners that is. I just don't like that some men suddenly think they're a little better than me. Some men are better than me, sure, but it isn't because they were lucky enough to knock the right rocks and find a vein. Some of those men think that because they have a few dollars, they are more important than others, or worthier or whatever. That's nonsense and I think you agree it is. I'd rather be down in the mine with my workmates, producing real wealth."

Lon reddened with anger and was going to say something, but Ben gestured not to be interrupted.

"The thing is, I like the work. And the people. Miners are the most optimistic, hard working and reliable people I have ever met. They—we—do battle with mother earth every day, and most days we win or at least fight her to a draw. And being among workers, I can hearing what is going on. Working with miners, I hear a lot. And, frankly, you are as honest as any miner—moreso than some. And I have to say, you give me good information from the other side of the fence.

"How kind of you to say so," Lon sarcastically chipped in.

"Well, it is important to know what is going on. And what you tell me helps."

Lon glared, silent.

Ben ignored the dig and continued. "You asked, Lon, so listen to my answer, alright? All in all, I have a pretty good feel for the life in the Cripple Creek Mining District. And sometimes I actually wouldn't mind being part owner of a mine. The thing is, today, in this camp, you have to choose. You can't be both, you have to be one or the other, miner or owner."

"I disagree, Ben. We're all looking for the same thing—wealth—and miners and owners are not enemies.

Ben shrugged, done with the subject. He turned to Abby. "I can't imagine that you take sides on this or any issue when talking with your students."

"No, I don't. And Lon knows that, don't you? I present all sides. Even so, I think the Union has the right to organize, to try to better their situation. That does not include the right to destroy others' property or to force men to join."

"Well, little..."Lon stopped himself. "Well, Abby, in fact union thugs do just that. And sooner or later the owners will have to put a stop to that kind of intimidation and vandalism. Maybe I shouldn't say this, but I know that Mr. Carleton, Mr. Moffat, and others have had conversations about this. A name has come up, I've heard it several times. Jim Warford. He has been known to knock a few union heads himself. Fight fire with fire and all that. I guess he'll be the MOA's head guy, their chief of security or head enforcer, whatever they call him. I am not sure if that is good or bad. Probably the latter."

Ben made sure to remember that name. "Let's hope things don't start to spiral downwards."

Abby knew there were no easy answers. "Enough education, politics and mining. Let's eat."

VIII

THE MEAL WAS ENJOYABLE BECAUSE THE INEVITABLE UNION discussion was over. Everyone was still smiling. The talk was of friends, growing up stories, national events, and books read. Local politics and affairs had been covered more than enough. Mostly they kept things light.

Ben's story was pretty typical of those who landed in the Colorado goldfields.

"I had to leave the Midwest. My brothers were taking over the family farm and business. The prospect of taking orders from customers or brothers held no allure. Do you have an older sibling, Lon? They love to order everyone around, is my experience. Me, I don't take ordering around very well."

Abby rolled her eyes. "No!"

"Anyway, I loved rocks and the outdoors. Not that it was easy to find rocks in Ohio, but I did. So it was natural to slide into a life of geology and mining. The right doors opened at the right time for me. Almost like it was meant to be."

Ben had never before been able to verbalize his dislike of being junior brother. He was a little surprised at such a concise description tumbling out of his mouth. After all that is what pointed him towards the outdoors and ultimately geology.

"So here I am in Cripple Creek Colorado, via a train trip, stops in Denver and at a number of other mining camps. Glad to be here!"

Lon revealed something, a close held boyhood secret. "When I was about thirteen I was afraid I'd get the Call."

This was news to Abby, who knew what he meant. Ben didn't. "The call? What's that?"

"That, my friend, is when God calls you for the ministry. You hear about people somehow getting that message. Folks give up medicine, law, farming, whatever, midstream. Apparently it is unmistakable if you hear it. At least that is what I understand from what I've been told."

With an impish smile he put his hands in prayer attitude and looked skyward.

"If God was talking to me he had a mouthful or something, 'cause I didn't hear it. Good thing! I sure see evidence of a grand plan out there, but I'm not to be one of them teaching it. Like you, being outdoors and knowing the earth spoke to me. So I too am here in Cripple Creek."

Abby was glad to see the common ground between them. They knew she taught because that was about all a respectable woman on her own could do. Sometimes she resented that lack of opportunity. She would have loved to be out walking the hills, looking for color. Or better yet, exploring parts of the world unknown to civilization. Still, she was happy to have a professional job and be out on her own.

Before the evening ended, the discussion veered back to local affairs. Over coffee, Ben again brought up current mining camp concerns.

"You know, Lon, I got to thinking about our earlier talk. When the owners whine about intimidation and so forth it wears thin. I mean, come on. Jim Warford on the MOA payroll? He is as leathery and tough an enforcer as anyone working with the WFM. It gets old for the miners who are happy with things as they are and don't like the Union. But the owners take it out on them as well as the union men. This whining about the WFM is like the lion whimpering in the jungle because the wildebeest it just killed is too small."

"Wildebeest? Ben, where did a three dollar a day miner learn about African wildlife?"

"I've not just been knocking rocks my entire life, Lon. I have traveled some. Not to Africa I admit, but I didn't just fall off the turnip wagon. Plus, I do read and observe." He smiled.

"But that is a story for another time. Like I said, the members and officers of the WFM are not intimidated. We all know that the Governor is, shall we say, sympathetic to them, the owners and their cause. Some would say he is in their pocket. Those rich men treat the State Militia like their private police force. So spare us the moaning about the money men feeling threatened."

He paused. "Do you remember the Pullman Strike, in Chicago in '93? Where the railroad men tried to refuse to handle Pullman cars until the owner of the Pullman car factory paid a fair wage? One of the first major work actions in the US. Do you know the government, the Federal Government took sides in that?"

"Well, sure, they were sympathetic to one side, the business owners."

"No, I mean the Attorney General, the chief cop of the United States, took sides. He specifically allowed and encouraged the railroad companies to hire and deputize anyone they wanted. So of course they hired thugs who had full law enforcement power. Those guys could stop people, search them, arrest whoever they wanted, the whole shebang. Tell me that is just and reasonable. You'd scream to high heavens if the WFM could hire and deputize anyone they wanted to. And well you should, in a fair world. But don't worry, Lon, it will never happen."

"I have to agree, that doesn't seem evenhanded. But that strike took place over ten years ago. Here, now, the Sheriff is a WFM man, or at least he is in their corner. The joke on the street is that he wouldn't prosecute a WFM man for a bludgeoning murder unless he was caught holding the bloody axe. And not even then if the victim was a mine owner! Now, you tell me, is that fair? What I'm getting at is, the owners have to balance the scales any way they can."

"Lon!" Abby was appalled. "The Sheriff is a fair and honest man. He enforces the laws evenly and fairly."

"Sure, and Ben here is the Pope. Seriously, what are the owners supposed to do when employees are beaten, or forced at gunpoint to quit a nonunion job? Are they supposed to just sit back and take it? We all know that Federation goons beat on and go after non union miners if they can get away with it. It has to stop."

Ben laughed, unamused. "And goons hired by the owners act like altar boys, right? Lon, you know that private security men beat on WFM miners every time they can get one alone. Please, let's at least be honest here."

"Honest? How about high grading? It is rampant. When miners steal ore and sell it on the sly, the owners lose thousands and thousands of dollars."

Ben couldn't help himself. "Poor boys. Losing thousands while making millions. It must be tough. Maybe we should take up a collection to help them can buy their caviar the days they come up short."

Abby rolled her eyes. "High grading isn't stealing. Not really. You can't steal something that is there for the taking. I believe the men who dig the ore should be able to cash in on some of it. Technically legal or not, it goes on in every mining camp. It has gone on since the first cavemen started working ore thousands of years ago."

She had the floor and wanted the guys to stop picking at each other so she kept at it.

"You know, don't you, Lon, there are crooked assayers who are happy to take the ore snuck out by the miners, right? They buy it at a discount, that's the way it works. Are they guilty of stealing too?"

He nodded silently and she continued.

"Well, you surely know that some of those assay labs are in fact owned or controlled by the mine owners. Those rich men know exactly what is going on. They know and they are complicit. The whole affair is a safety valve. It is a big game for everyone to make money on the side and let miners get away with sticking it to the man. Everyone pretends it is on the sly and no one else knows. All the players are winking and nodding at each other. Don't you complain to me about high grading."

Ben: "I have friends who high grade for a living. It is a well-developed part of the scene here. There are plenty of men and labs involved. Like Ab says, it has gone on since Adam and Eve."

Lon: "The thing that gets me is many, maybe most, of the miners seem happy with their pay and conditions. The high graders are in it for the ease and the thrill, and they make money in the bargain. But the typical miner doesn't have to resort to such acts to earn a good living. I'm told that most miners think that Big Bill Haywood is at least as interested in power as their working conditions. That comes first and the welfare of the miner in the hole comes after that."

"Frankly Lon, sometimes I struggle with that myself. I guess as a WFM member I at least know what the local officers are up to and where the union is going. I'd rather that than be a non member and be surprised by their actions, wholly at the mercy of events. Not to mention the heat the union enforcers give the non union men. That is especially true if they try to call a strike. I sure wouldn't want to be surprised then to get jerked around. Or beaten."

"Let's hope Haywood doesn't do that. Call for a strike."

Silence, all in agreement but no one knowing what to say. Ben jumped back to other local issues.

"You know, another part of the picture here which doesn't get much attention is hauling. Of course railroad tracks aren't laid to every mine. Getting ore to the surface is just half the battle. Somehow it has to get processed, refined. Needs to go from the mine to the railhead for shipment to the mill. There's money to be made there, filling that gap."

"Yes, there is." Lon smiled. "Ask Bert Carleton. He has a near stranglehold on hauling ore around here, at least servicing the big operations. And he is doing very well, thank you. You know that he would take a dim view of new competition."

"At the Emma May, like I said, for a while it looked like we had the big vein. Of course we had to move ore from there to be processed. Carleton's men acted like they were doing us a favor to send a wagon around. We almost had to get in line and beg just to get them to come and take our ore. It was almost like we were too small for them to bother."

Abby smiled. "I think the saying is, 'Them that has, gets; them that don't, don't.'"

Ben nodded and so did Lon.

"We had to pay for them to carry it off, but not too much. We were small fry which is why we sometimes had to wait. I know the big boys pay lots of money and keep hauling crews busy. On top of that, some of the wagons and cars need armed guards. The ore is so rich they have to make sure people don't grab it on the way to the smelter."

Lon smirked. "Speaking of high grading..."

Ben smiled at that. "But I digress. The big boys pay a lot, mines like the Independence, the Portland, Vindicator, and others. But if you add up the tonnage of ore moved by the small guys like we were, I bet it would be almost as much as what is produced by the top big mines. Or at least it adds up to enough to be worth hauling. It is just in relatively small amounts is all."

"So what I am getting at is, maybe there is money to be made in short hauls from the small mines. Specialize in them, take the business Carleton is too busy to bother with. The owner of one wagonload of ore ought to be able to send it to the mills as reliably as the Independence Mine sends its carloads. Right now it goes mostly to the mills in Colorado City. Some of it goes the other direction, down to Canon City."

Lon considered what Ben had just said and added a thought. "I think it makes sense to watch what the small guys do and how they do it for a while. Maybe you're right, Ben. Maybe there is an opportunity here."

"Another thing. The high grade stuff, the ore which assays at many ounces per ton, is easy enough to process. But many mines now are bringing up ore which assays less rich. Still it is worth mining but more ore has to be put through the works to get the same amount of gold, silver, nickel and lead. New processes are being invented for that. And those mills are less complicated and expensive, more efficient. They are being built right here in Cripple and Victor. Much of that less rich ore will be worked or at least concentrated here."

"So hauling wagonloads to a local smelter might be a winner for someone." Lon again said, "Now you got me interested. I definitely want to watch how this develops."

Abby watched them talk, eyes back and forth as if following a tennis match.

"This is something I love about this gold camp. Listen to you guys, a wage miner and an owner. Many would think you are enemies. It is pure Cripple Creek: you're talking ways to help the little guy and make money in the bargain. That is if you don't make it big on your own. Ben of course, wants to stay in the mines moving ore and listening to his fellow workers." She paused and smiled.

"Ben, are you going to become a nabob in the freight business? Paying others to get their hands dirty while you arrange and count profits? Or are you going to keep moving ore in the mines?"

Lon laughed. "It'd be crazy to do that. Stay in the mines, that is. Why do hard work yourself when you can organize others to do it for you? And make more money in the bargain? I ask you, Ben?"

The bald statement caught him out. Ben had never considered his work in that light—doing hard stuff or organizing others to do it.

"You know, I'm not sure."

Lon glanced at Abby. "This idea might have legs, Ben. Hauling lots of small loads from the smaller mines. It might be a chance for us to make it big. Surely you can see that?"

"Maybe so. But I'm not sure I want to make it big. That brings its own problems and unpleasantries, I'm sure. Say we hit it big. If you or I strike the big one, we'll have 'friends' coming out of the woodwork, trying to take our money. And half the town would be jealous and angry at us. Who needs that? Maybe I'm happy to be a miner, working with straight, honest people. Looking after myself only with no need to be responsible for others."

He grinned. "It is easy to get a good night of sleep when all you have to worry about is your own work. But a nabob, a nabob's day is never over. You are always having to worry about others. Did they do the job right? Did they steal? Did they lock the gate on the way out? Will someone sue me for some bogus reason? Or worse, a real reason? And then there's your money, where is it, who has their hands on it, and so on. Me, when my day is done, I can forget the job and go do what I like."

He looked at Abby. "Right now what I like to do is hang around with the local teacher."

LATER, BEN HEADED HOME. NOT TO THE HOUSE HE COULD have regained when Lon bought out the Emma May. After some soul searching, he let that house go. He figured he didn't need the expense and responsibility. Now, he lived in a rooming house. It was cheap and easy, and he could be sociable or not depending on how he felt and what else he had to do. Plus, his only obligation was to pay the monthly bill. He could concentrate on the important things in his life: gathering information on local labor issues, the actions of the mine owners, reporting to Big Bill Haywood, and Abigail Bosini.

As he walked, the face of a new man at the rooming house came to mind. He seemed a nice guy. The man didn't seem to be wild about the union. Nor was he an independent, anti union firebrand. Really, he didn't say much one way or the other when it came to labor issues. George or Joe or something. Last name Harris or Henderson, starts with an 'H.' He recently came in from Colorado Springs, looking to make his fortune like everyone else. The thought faded as he idly watched men and a few women. Some walked his direction, some towards him.

Not long and he ran into some old friends. One of them looked worried, jumpy and underweight, the other was as big and stolid as ever.

"Hello Owens. Hey, Juni. Haven't seen you guys in a while. What are my old trail pals doing these days? Still avoiding the mines is my bet. But you look good, both of you. Juni, I see your shoulder seems to be healed and healthy. What's up?"

Arms atwirl and suddenly animated, Owens grinned. Juni remained two or three feet back, strangely standoffish. He said nothing, just watched as Owens talked.

"Hi Ben! Yeah, long time no see, say, I hear you are back on shift work, and have an in with WFM, and someone said that you are hanging around the school a lot, sweet on the teacher, any truth to those rumors?"

"I'm not sure about all that, Owe. That covered a lot of ground! Well, yeah, it is true that I am back at three dollars for a full day down the hole. I have to say it is good to be back. What's that you say about the union? Yeah, I do go listen to Big Bill Haywood when he's in town but that's about it." Ben downplayed his knowledge and role. In Cripple Creek one went easy in public when the subject was labor or owners.

"And the girl?"

He reddened a little. "I have made the acquaintance of the teacher, yes, and we seem to get along."

Actually, Owens was paying little attention to his responses. The man easily got worked up and into his own zone.

"I also heard a rumor you own a small part of a mine here somewhere, how does that work with the WFM?"

Ben glanced at Juni and smiled. No response from the big man. He didn't really answer this question either. No reason to get into all the sticky details, especially with Owe. He wouldn't listen or care in any case. Ben simply sidestepped it.

"Hell, Owe, half the miners in Cripple own part of a claim somewhere. So what?"

"Just wondered." He looked around, settling down, his arms now loose at his sides. He was obviously making sure no one could eavesdrop. His voice, no longer loud and brash, was low. He spoke quietly, so softly that Ben leaned in to hear.

"Ben. I just wanted to check if you've become a hoity toity owner or are still a working man and it sounds like you are still a good solid miner and more interested in the workers than making money off their sweat, glad to see that. Anyway, I gotta tell you, things are coming to a head and I warn you to look for a surprise for the scabs at one of the big mines—they will be sorry."

Owens may as well have hit Ben up side the head with a lake trout. The surprise was complete. "What? Sorry? Who? Sorry about what? What are you telling me, Owe?"

He clammed up, at least as much as Owens Anderson could go quiet. He looked down, kicked dirt, stuck hands in the pockets, stood still which was rare for him.

"I shouldna said nothin'. You didn't hear a thing from me, okay, and just forget you even saw me out here, you take care of yourself Ben, keep your head down and be careful."

He turned and walked, practically ran, away. Juni remained, stolid and unmoving.

Ben shook his head. He reviewed what he knew of upcoming WFM actions, and none of them remotely fit Owens' description. For a moment he watched Owens scurry off like a pig chasing the slop bucket. He turned to the other man.

"What was that about, Juni? Things coming to a head at a big mine and scabs will be sorry? Has he been visiting Papa San's opium den? Or is this for real? Is someone planning something? Where? When? What is going on, my friend?"

"I don't know, Ben. No one tells me anything. But..."

"What? But what, Juni?"

"Well, like Owe said, you didn't hear it from me. Right?"

He said nothing more, just gazed at Ben, who nodded.

"Yeah, sure, I sometimes overhear things in a saloon. Or in line at work, or something. Whatever."

With a thin smile, Juni spoke. "You remember I told you about that guy from Idaho, the one helping me with high grading?"

"Sure, Herman or Herb or something like that, yeah I remember."

"Don't say his name." Juni hissed, abrupt and almost fierce, then continued after a moment.

"Here's a 'what if' for you. Say there is this guy I might have seen around. A guy from another state. Well, what if this guy has a temper and holds a grudge like an ape grips a banana. And he likes kaboom. Really enjoys it when he can blow something up. He's good with dynamite and loves to use it, rig it up to fire just right and so forth."

"So? Juni, this description fits more than a few miners I know."

"Yeah, well, on top of all that, what if this guy is angry, has a temper like I said. What if this man hates owners. Hates 'em bad. But he really, really has a case for strike breakers. The guy might even call them scabs. And what if there are people or a person who might go and take things into their own hands, know what I mean? And maybe some other folks kind of get swept along, especially people with a lot of nervous energy who talk before they think."

He looked away, shuffled his feet, thinking, and made a decision. "Listen, Ben, I gotta go. Like I said, you didn't hear anything here tonight. Right?" He too turned and left, ambled away a different direction than Owens was last seen.

"Wait, Juni…" Ben watched the big man diminish and vanish in the dark.

BEN HUSTLED TO THE UNION HALL. HE HOPED TO TALK TO Shuler or even Haywood if he was around. He was in luck.

"Good to see you two. We need to talk."

Haywood gave him the up and down, clearly not happy at being interrupted. He felt half naked after that and was glad he wasn't a woman. "What's up, McNall?"

"Well, you told me to keep an ear to the ground. I hear all kinds of stuff, much of it clearly just gossip. Junk. But sometimes it has legs. My rule is, if I hear the same info from several places, maybe there's something to it. Have lately heard a disturbing story."

Ben looked between Shuller and Haywood, who were both focused on him.

"Supposedly someone has or will stage a dramatic demonstration. I hear one is being planned, something to get at scabs. Is there anything I should pursue, or know about this?"

"No. Nothing on our end, not this local or any that I know of." Shuler rearranged a stack of files. "Of course, we encourage members to read friendly newspapers. We don't want them looking to the garbage papers that apologize for the owners. And we encourage men to join. But no, other than the usual 'persuasion' we are not planning anything big. Tell me about what you hear."

"Well, nothing precise or that I can nail down. Just that someone who likes dynamite wants to blow up a bunch of scabs. Kind of vague and threatening, nothing as to timing or where or anything like that."

The two union men exchanged glances, and Haywood reddened. He acknowledged and dismissed Ben in two sentences.

"No, nothing that I know of. Thanks for the heads up, and let us know if you hear more."

After Ben left, the big man drummed his fingers a moment, then spoke.

"I want you to keep tabs on this, Frank. But I don't want to know too much, if you get my drift. And you ought to take the same tack—we want to know but we don't, right? We have to keep our skirts, or I guess our shirts, clean. The union does not want to know any more. But I'll say this: If someone were to smear a few scabs or managers or owners over the mountainside, who cares? I don't want the WFM doing such things. I don't want any connection with that kind of stuff. But…it wouldn't be a bad thing. The sooner the capitalists are eliminated, the better."

BEN DID NOT HEAD HOME FROM THE UNION HALL, TO HIS SAFE anonymous rooming house. The whole scenario troubled him. Owens and Juni weren't alarmists nor would they snitch lightly. But something was in play. Whatever they described was likely coming down the pike, maybe at top speed. He needed to talk, so veered to familiar safe territory.

Before knocking on the door he took a deep breath.

"Abby, sorry to come back. I imagine you just saw Lon off and are ready to call it a night."

"Well, yes. I was. But I'm always glad to see you. Are you alright? What brings you back?"

"Well, I came across a secret. But the way I was told makes me think it shouldn't be a secret."

"Oh?"

"Yeah. I ran into some old friends, guys I came to town with years ago. We've gone our separate ways. I don't think I've even mentioned them to you, it is that far back. Neither of them is working in the mines any more. One of them is on the…I guess it doesn't matter what either of them is doing."

"Are you sure?"

"Yeah. The thing is, they each, separately, warned me of something coming. Some action, maybe violent, at a big mine that will supposedly involve, get at, non union workers. Lon would call 'em strike breakers and WFM calls them scabs.

Whatever, they are people and I don't know what may be happening. Anyway, my pals wouldn't be more specific about the threat."

"Do you believe them? Maybe they aren't specific because they don't know. Lots of rumors roll around out there, Ben. Or maybe, if they are in on things, they do know but are afraid to say more. What, are they blowing smoke or is it real?"

"That's the question, isn't it? What do you think I should do?"

It seemed simple to Abby. "Try to find out. If you can't run the rumor down, mention it to Shuler or Big Bill. Their reaction will tell you something. If they know all or something about it. And whether word should be on the street. Or maybe it is somebody else making trouble for the union and the owners."

"Haywood? I did, just came from there. I asked him generally about it. He denied any plans or knowledge of anything to harm people. Said the union doesn't do stuff like that. Shuler agreed. I hope this doesn't give them any ideas. I don't want him to expect me to run down every leak or rumor. Half of what I hear is told me on the promise to keep the source private. Certainly, the men who talked to me on this demanded anonymity. I promised and have to honor that. Jeez, what a circus."

"I guess you can wait and see. Maybe this is all rumor."

"Maybe it is. But Idaho Springs had an explosion a while back, at the Sun and Moon mine. Someone blew up the power house there. The union got blamed and union men got run out of town. Maybe that is what is going on? Is this someone trying to make trouble for the workers, and the union?"

He shrugged. "Augh! This is bad. I need to sleep on this one, Abby. Thanks for talking it over. And keep this under your bonnet, please? Especially from Lon, at least until we figure it out."

She paused, thought. "Alright. For a little while, I'll keep quiet. But if we find it is real, we need to warn people. Doesn't matter who is behind it, we don't want people hurt."

"Oh? And just who do we warn? And how?"

"I don't know. And we can't do anything now. Go home and get a good night's sleep, Ben. Will I see you tomorrow?"

"Long shift tomorrow. It'd be late by the time I could come by. Let's just meet up at the union hall in the evening, day after tomorrow. Good night."

AFTER HIS SHIFT TWO DAYS LATER, BEN WENT FROM THE LIFT station direct to the union hall. Abby walked from the other direction and they met in front.

"Big Bill is back in town. I think I told you I saw and talked to him the other day. And you no doubt remember, he got his wish. He is king of the world now, at least the WFM world. Members don't have to, or get to, vote on matters any more. As good Union men they now have to take direction from the local officers. Isn't that the reason people left Russia and Germany and all, to get away from one man telling them what to do?" Ben shrugged.

"Be that as it may...Those union officers telling the members what to do get their orders straight from Big Bill. We'll see what instructions he has for us peons tonight." He started to enter the hall.

Abby touched his shoulder, stopping him. "Have you heard any more rumors?"

"No, nothing. Maybe, and I hope, it was just talk or someone's fantasy."

They went in, sat near the back, and listened to Frank Shuler. He went through agenda items such as the schedule for an upcoming Auxiliary bake sale, the budget, and the decision to paint the interior of the hall. The few members in the audience struggled to stay awake. Then he switched to a list of mines that blackballed WFM members.

"Listen up now. This is important! The way it is going now, a man needs to be on Clarence Hamlin's list. He needs a card from the MOA—be on Hamlin's list—in order to get hired on at these mines."

He waved the paper list. "And he has to certify he is not a WFM man in order to get that card. Can you imagine, making a man do that just for a job? Isn't that why people leave Russia or Germany, to gain freedom to work where they want? We need to break that system."

He looked over at the Executive Secretary of the Western Federation of Miners, nodded, and sat down.

Big Bill took charge.

"Union members have shown their wisdom yet again. The change in procedure I have been discussing with you all has been approved and is working well! Thanks to you for giving more authority to your officers. Now that they have broad ability to act, we have moved forward. We are growing in membership. And we are gaining momentum! We will gain the right to have our members work at any mine they want to, you wait and see. One other item: this is just for our members to know, and is not to be repeated. We have men in some of those 'blackball' mines. Undercover."

He looked around. "Now, no one, not a soul, but us people in this room know that. So if word gets out, we will know who to come talk to about it. Am I clear?" He didn't say more, just looked around again, then went on.

"Our men in those mines say the workers there aren't happy. That many of the miners would be happy to join the WFM if they could. At some point we will need to get those mines turned around and on the WFM bandwagon. But not yet. We have things to do."

He paused, took a drink of water. As he spoke he glanced at Ben.

"First, we will take action to make non union miners regret that they didn't join us when offered. Those boys are no better than a scab on a wound. Those non union workers, those scabs will be made to think twice before going to work for a greedy owner. An owner who doesn't give a damn about the workers."

He stopped, glared out at the audience. "I want you to read the Union news and friendly papers. Only. Do not, do not, read the rags sympathetic to the owners. They will poison your mind with lies. We all need to keep our minds sharp in this battle."

Then the spell broke, and he strode back and forth across the stage. "We, the Western Federation of Miners, do give a damn about the workers! We stand for a living wage and for a safe mine! We are for everyone to share in the wealth we produce, not just a few."

He started to pace agitatedly, then stopped and gathered himself before going on.

"And we will do our damndest to discourage any man from working in a non union mine! Those men are sucking the life out of the community! They will be discouraged, you wait and see. Just wait. Now, we need you to go to work, do a good job. And after work, be a good citizen—no fighting or name calling or any of that. If things need to get tough leave it to us. The WFM will take care of you and the community. No freelance action, you hear me? Meeting is adjourned, go out there and make the WFM proud!"

Big Bill talked a good line, but like any senior executive, he knew there were some things he was better off not knowing. He went to great lengths to remain in the dark about some plans that were being considered.

OWENS HAD TOLD BEN AND OTHERS ABOUT THE FRIEND AND mentor he had fallen in with. Actually he met the man through Juni, who was off doing his own thing now. Owens' new sidekick went by the name of Harry Orchard. Owens was pretty sure that wasn't his real name but didn't ask. Cripple had many people who had reinvented themselves when they landed. It was poor form, actually not a good idea, to ask too much about a person's past. They would share what they wanted to, and some things you just plain didn't need to know.

Orchard, Owens learned, had drifted to their camp at the foot of Pikes Peak from other mining settlements. He boasted of being run out of Idaho for union agitation, in the Coeur d'Alene mines. The man had a temper and it seemed to Owens that a lot of his actions were violent. Maybe not violent all the time, but angry and definitely not peacable. Owe now thought of the man as being out there on the edge of the law.

Coming to Colorado, Orchard went straight, kind of, for a while. Harry gravitated to high grading where he met Juni.

Both men were happy with the lifestyle. Juni liked the good income, free time, and being his own boss. But he had no stomach for the active anti-owner work which Harry talked about more and more. Somewhere along the way he had introduced Harry to Owens. They got on well and Owe ate it all up, the high grading and the anti-owner work. Juni was glad to get away from the unsociable activist.

The two mischief makers, Orchard and Anderson, got along well. So well that Harry proposed and Owe accepted going on a mission, an adventure. It was night and they were in the bowels of the Vindicator Mine. This was a rich and well respected operation in town, one of the richest. Also it was one which the MOA controlled. Its workers were all on Mister Hamlin's list and were definitely non WFM.

Late that night they trespassed onto the Vindicator Mine. Coming in via an old shaft with rickety access, they were quiet and intent. They weren't there to grab ore and run. No high grading for them tonight. Harry had other fish to fry.

Owens was only half listening at first but came around fast.

"So we'll go down to the seventh level. That level is being actively worked, ore being moved. Taken out by scabs!" The tone of Orchard's voice chilled Owens. He himself was usually talking and so he was seldom really aware of voice tone and intent. But tonight he listened, and something about how Orchard spoke made him pay extra attention. He fought to be still, keeping his arms from their usual mantis like motions.

"Why go there, clear down to the seventh? God, that has to be hundreds, maybe a thousand feet down. What will we do there?"

"You ask too many questions, Owens. See that car full of dynamite? Go grab a box, a full one—careful like—and come on. It only weighs fifteen or twenty pounds so you can carry it alright. Be ginger with it, don't drop it whatever happens."

It was after hours. The normal entrance was bypassed and they came in via an old and unused route. Also they took detours to avoid anyone and not leave obvious tracks. It was a maze but Orchard was confidently leading the way. Finally he stopped.

He spoke just above a whisper, direct into his pal's ear. "Here we are. I think this is the seventh level. So we rig it up."

"Rig what?" Owens couldn't help himself, he spoke loudly.

"Jesus, Owens, be quiet, stop squirming and fidgeting, and shut the hell up with your questions."

He pointed.

"You just go set that box in front of the lift, where the gate opens, and I'll rig this pistol to shoot and set it off. When a carful of scabs comes here, kaboom! No more scabs. What we want to do is make men stop and think. It will make non union men think twice about what they're doing. And it will rid the world of a carful of vermin."

"Alright, alright. I just wondered what you wanted and now I know and I'm doing it and do you think this is really a good idea, Harry?"

"Hell yes. A quick death is too good for scabs. They're just traitors to the working man. Now get to it."

Owens quickly and a little fearfully did as told. As they left, each looked back, one with dread and the other with satisfaction. The setup would work perfectly, blowing up when the lift door opened. Neither man realized they were mistaken. The booby trap was rigged on the sixth level, not the seventh. The sixth was not active. It was only checked, visited occasionally by a foreman or supervisor.

So nothing happened for a time. Orchard wondered if the trap was discovered. It was not. He soon saw to his delight that it worked. Perhaps not exactly as intended but it did cause a big boom.

———

THIS MIDNIGHT FORAY ACTUALLY TOOK PLACE A DAY OR TWO before the union meeting Ben and Abby went to. After that meeting, the couple strolled and talked. Ben detoured his way in order to stay with her as long as he could.

"Ben, whatever Haywood told you, it sounded to me like something is being planned. By or with the connivance of the WFM. Even though they'll set it up to be able to deny it all. If there is something done, that is. The thing is, maybe your friends were right."

"I can't disagree, Abby. But what do we do with an anonymous, fuzzy prediction of 'something'?"

"Well, we should tell Lon. Would you talk with him, or do you want me to?"

"I'll talk to him. But no names—I won't give up my friends."

"Tomorrow night, alright? My place? After work?"

They stood, holding hands, gazing into the other's eyes. This was as much affection as it was proper to display in public.

"Alright. See you then."

———

NEXT NIGHT BEN STRODE FROM THE LIFT DIRECT TO ABBY'S, not stopping or talking with anyone. He was hot and tired. He would rather have been cleaned up, but no time. He focused on what he intended to say to Lon. People in the street seemed subdued and upset but he paid little attention.

Lon was there, holding a newspaper, and he too was upset. Abby was looking out the window, dried tears on her face.

He shook the paper at Ben. "Did you hear?"

"No. Just off work, haven't talked to anyone. What?"

"The Vindicator. Explosion. Killed a foreman and the superintendent. When they got off the lift for a routine check at the sixth level."

"My God. What happened? I mean, do we know how there was an explosion? What set it off, I mean?"

"Apparently it was set. Someone intended to kill people."

Ben sat, trying to grasp it all. He looked at Abby. "Did you tell him?"

"No."

"Tell me what?"

"Lon. About two weeks ago—after we last had dinner—I ran across some rumors. You know how people talk. Anyway, one rumor I got then was, at a big mine with non union men 'something' would happen. That's all I could get, 'something.' The other rumor kind of tied in. It was that there would be an explosion involving non union men. Nothing happened so I kind of let it go. I figured that you can't believe ninety percent of what you hear on the street. But. Now this."

"Who, Ben? Who told you? Why didn't you say something?"

"I can't tell you who, Lon. Even if I was sure." He lied about not knowing who, not sure why, then it came to him.

"If I were to tell you more they would probably end up in a kangaroo court and get shot. Me too, probably. The WFM plays rough. Plus, I may learn more from these sources. If something more is in the works, God forbid."

"I suppose. But damn it, next time, if there is a next time, tell me. Lives are at stake! Please. I'll guard your privacy but maybe we can avoid bloodshed. More bloodshed."

He pivoted, staring at his sister. "Did you know about this, Abby?"

"Yes, Lon I did. But like Ben said, nothing happened and it was all so fuzzy that we decided it was probably all rot. So we didn't tell you or anyone."

Lon was hurt but knew if forced to choose between him and Ben, Abby would not go with her brother.

"Well, if you hear any more, don't sit on it please." He sighed, tossed the paper down, and rubbed his temples.

"This will bring back the Militia, you watch. Just when things were settling down and we could all go about our lives. Men wanting to better their lot is fine, but killing to do it is not! The WFM has crossed the line."

"Do we know WFM did it?" Abby's voice was small but firm.

"Nah, they'll deny it. And have an alibi for everyone. Still, I'm sure they did it. Or a WFM man was encouraged to do it on his own. I'd bet my last dollar that it was done with a wink and a nod from them!"

IX

BEN SHOOK HIS HEAD AT LON.

"The union will rightly deny it, Lon. I'm sure they didn't do it. They are not in the business of killing opponents. Heavy persuasion, yeah, I guess, but murder, no. This has to be the work of a madman. Or a provocateur. Either way, any thinking man can see how things came to a bombing."

"Things came to a bombing because the miners union is…"

Abby was fuming. "Oh, come on all, Lon. The Mine Owners Association, the so called MOA, has been waging war on the miners union. They hired that Jim Warford to be a surrogate Sheriff and knock heads. I agree, mine managers shouldn't be killed. But the MOA sure tries to make it hard on the union."

"But the MOA doesn't rig bombs to kill Big Bill Haywood when he comes to town, do they? You can't equate giving the opposition a hard time, which the MOA does, with killing innocent men doing their jobs."

She didn't talk over her brother nor did she acknowledge him.

"And they have Clarence Hamlin, Secretary of the MOA. Secretary my eye. He is the man who will blacklist any union member. Hamlin and his list! The whole system is devilishly efficient with his work permits. He issues them only to men who will quit or repudiate the WFM. He and his rich friends, Penrose and Tutt among others, have made it hard for Federation members to get work. Other than tending bar or mucking stables, that is. They can't be hired here in the district if they don't have a permit. At least not as a miner."

Ben: "If the owners get their way a WFM man won't get a job in any mine in the state, even the region. They are building a list, and they talk to other owners around, sharing names of 'troublemakers."

"Well, Haywood and others of the WFM would destroy the owners if they could. I mean, these men, the owners have their homes and wealth, their very lives at risk. Don't they have the right to defend themselves? And then the oh so innocent union allows, even stages, acts of terrorism. Beatings, blowing up trains."

"What else could they do, Lon? Go hat in hand to the MOA?"

"What they got, Ben, is the answer they deserved for the beatings and vandalism. Last time this sort of thuggery got out of hand was back in 1894. About ten years back. Then, the militia came to town. Sherman Bell, the adjutant of the state militia came and got things under control."

"Yeah, Teddy Roosevelt's sidekick, one of his Rough Riders. Really likes himself. Some guy."

Abby rolled her eyes. "You know, I heard when he came to town he declared war. She rustled a stack of newspapers, pulled one out. Before she read from it she explained.

"I have these old papers for a research project, and just saw the article a day or two ago. Back then he said, and I quote: 'I'm here to do up this damned anarchistic Western Federation.'"

She rolled her eyes in outrage. "' I'm here,' he says. To 'do up' the union."

There was fire in her eyes. Ben had to tear himself away from them back to Sheriff Bell.

"Can you imagine that, the arrogance? I, the great General Bell? This self important ape wants to 'do up' an association of free and willing men? Who the...Who on earth does he think he is?"

Lon thought back. "You need to remember this all happened about the time of the Pullman Strike. Back in Chicago. Labor was rising everywhere and the government actually sent soldiers to try to move trains. Workers refused to handle any train with a Pullman car, in an effort to get Mister Pullman to pay reasonable wages to his workers. The soldiers also tried to keep the peace. In a few cases, soldiers fired on civilians and it was damn near civil war. Everyone, everything was becoming heated. No one knew where it was heading. Our governor sending the militia in here was not out of line given the times."

Ben stood and paced. "Maybe it was an inflamed time. Still, the government warred on the people. General Bell promised it and war he gave us."

Lon glanced at him then Abby, noticing he used 'us' not 'them.' Abby's expression told him she agreed with Ben. Ben went on.

"Not ten years ago here in Cripple. Bell stationed troops all over the district, put up searchlights on hills, and set up the bull pen! Some damned open air corral guarded by soldiers, to hold the ferocious, beastie labor men. All that those wild men were after was a fair wage, not jail."

Abby seized on that. "The bullpen. Ah yes, mister modesty Sherman Bell had WFM officers rounded up. Frank Shuler was one of them. Altogether four of them got tossed into that outdoor jail up in Goldfield. In the United States of America, rounded up like cattle and publicly jailed! No arrest warrants, no order from a judge. They were just rounded up, marched down the street and fenced in. And then Bell made an attack on the free press!"

Lon laughed. "Bell got his comeuppance on that. Makes a good story now and I guess it was a good story then! He sent in forty five armed militiamen. To take over a local newspaper, the Victor Record. And to cow the staff. He figured it'd take that many to corral four employees and the editor of a small town newspaper. I guess the Record is sympathetic to the union cause. I don't read it, never have. But it did no harm, and was just exercising free speech and all. Anyway, Bell's forty five soldiers marched away with their five newspapermen. He figured that was that, did Bell. The good people of Victor had other ideas."

Abby grinned. "I've heard stories. He got the men but didn't figure on the newspaper woman. They left one woman there, no doubt figuring she would just sit, weep and wring her hands. Not so much! My friend Emma Langdon, a linotypist for the paper, took things into her own hands. Literally. She was there at the Record when the men were marched away. Rather than cower, she barricaded herself in. She defied Bell and the soldiers to invade the privacy of a woman. Even Bell wouldn't, didn't do that."

"Ah the power of an angry woman..." Ben couldn't help but interject, eyes alight.

She went on. "You got that right! Emma stayed in all night, setting up the paper, which she published next morning.

She actually set the type and ran the press, all alone. I love her sense of humor and defiance. She set up a banner headline. It is classic and she must have given it a lot of thought. I will remember this to the day I die, it read: 'SOMEWHAT DISFIGURED BUT STILL IN THE RING!' And the best part of it, she then took copies of the paper to the bullpen in Goldfield and handed them out. Especially to the soldiers there. They didn't expect that!"

"No they didn't did they." Lon smiled. "And that was just the first reversal General Bell suffered here, a decade back."

Ben agreed. "Thank God for the judges. A judge ordered Bell to produce the four imprisoned union men. He marched them under armed guard to the Court House. Set up a Gatling gun, had sharpshooters stationed everywhere. But the judge wasn't impressed. He heard the case, listening to both sides. Then he ordered the release of those four men. But first he reprimanded Bell for taking the law into his own hands."

Abby couldn't help herself. "Yeah, Bell harrumphed and moaned and rattled his sword for a while, but released the men within a day. The big tough General had no choice but to submit. Like you say, Ben, thank God for judges and the law."

Lon sighed. "And that should have been that. Most people were ready to get on with life. They just wanted to go to work, come home, enjoy the family, maybe save a little money. Everyone was tired of the upset and the drama and the bloodshed. Many were happy that Hamlin's work permit system took the wind out of the WFM's sails. Men had jobs and were getting paid a pretty good wage. But Haywood wouldn't give up. He had his goons beat or kill non union workers, blow things up, and generally make life miserable until he got his way."

"He didn't do that," Ben protested.

"Well, Ben, Haywood himself didn't do any dirty work, I'm sure. But he had to know it went on. He's the Executive Secretary, right? He could have shown some leadership. But he did not criticize nor did he stop it. Truth is, he approved of the actions."

Lon was just getting started.

"He is at work on many fronts. Not only here has he made mischief, but he has stirred things up in other gold camps as well. His WFM is kicking over the beehive in Telluride, Idaho Springs, Central City and others. He is a troublemaker. Just think about this Vindicator business we just heard about. You don't allow innocent men to be assassinated—and make no mistake, this was just that. The union may as well have pulled a trigger on this. They are out of control. If they can't be reasonable over there at the union, maybe the whole thing has to go."

Abby and Ben looked at him.

"What? Do you want to encourage killers to run around and blow up other people and their property?"

A pause, then Abby. "Well, brother, they have no choice."

"No choice? There are scores of mines here in the district looking for men. Any man willing to work can get on and do well. Why do some thugs have to be violent and destructive?"

Ben glanced at Abby and there was some communication, some spark between them. He answered Lon's question.

"We don't want to encourage such awful behavior. It has to stop. But at the same time, the owners' hands aren't clean. We have Haywood and his hired fists. They have Jim Warford and their own rented thugs."

"Yeah, so?"

"The thing is Lon, we have lawless owners, or I should say, owners who think they are the law. Unfortunately they often manipulate the courts to be on their side and the little man has few choices to fight back. I say, we say, a rich man should not be able to use and discard a worker like an old rag." As he said this he looked at Abby who nodded in agreement.

"And we say that workers should be paid fairly for all the work they do. Workers should share in the riches they help produce. And there should be no penalty if a man—or a woman—wants to join a union."

Lon looked at his sister and her friend. 'Friend,' he thought bitterly. His sister was lining up with a union goon.

He straightened. "Well, I see little value in the union, at least any union that does what the WFM is doing now. And a fair person has to question the logic and motives of anyone involved."

He paused, unsure, then decided. "It is getting late. I had better go." And he walked out.

"Lon!" Abby followed him to the door but he strode away saying nothing.

She put hands to her face and sobbed. Ben put an arm around her shoulder.

THE NEXT DAYS DILLON BOSINI WAS BUSY. HIS ATTENTION and energy went into the Double I. The mine was progressing, was even paying. The profit was not headline making but a profit it was. The vein seemed to be holding true and the assay reports were good. He worked to grow the business, concentrating on that and not on thinking of his sister.

His time was not spent looking over the foreman's shoulder but just being visible. Plus he wanted to find a way to pay his men more. Ben, or was it Abby, had a point. Workers should be paid for all the hours spent at the mine. He was too small to buck the big owners. He knew that if he upped the wages singlehandedly there would be pressure from the MOA.

The Double I was not a top tier mine but still…Of course he was wary of claim jumpers, those who would try to take the mine. Also, if someone could set a trap in the Vindicator they could easily do so in a small place like the Double I. Everyone on the crew paid attention to strangers in the area. And they double checked the rigging and setup every time they entered the place.

Time went by. He read the papers with interest. Sure enough, the assassination of the men at the Vindicator brought back the Militia and General Bell. Bell threw his weight around again, rounding up union men and stuffing them into a 'bullpen.' There were beatings. You didn't need to be a union man for this. This was a time which allowed, even encouraged 'getting even.' The General outlawed street meetings, loosely defined.

Lon, like many Cripple Creekers, was of two minds about all this. On the one hand, it was hard not to feel that 'If they blow things up, a few fists serve them right.' But on the other hand, no street meetings? For Bell and his men, that meant that three friends couldn't talk if they happened to meet out in public. A rifle carrying soldier would come along and break up their 'street meeting.' Not good.

Across town, in the union hall, Big Bill Haywood was working the situation. This heavy handedness was easy for him to exploit. He made sure that Union owned papers and other sympathetic reporters told of Bell's excesses. He did his best to bring them to everyone's attention.

The news was as bloody and strife ridden as ever. Of course people knew about Cripple's labor struggles which were mirrored by goings on in Telluride, Central City, and other camps. Also many events elsewhere took attention: David Moffat's railroad from Denver to Salt Lake had crossed the Continental Divide; the Russo-Japanese War was in full flame; the Wright Brothers their flyer held peoples' attention; American troops fought on in the Philippines and Cuba, there was unrest in the Tsar's Russia...With all that was going on, the papers made for lurid reading.

Abby and Ben were sitting, reading, one evening. Abby shook the paper in disbelief.

"Ben, did you see this article? On the confession of a so called terrorist? Right here in town?"

"No, haven't seen that section. Tell me."

"It quotes a 'union man.' It isn't clear if he is a staff member or a card carrying miner. Or a hired hand, one of Haywood's hired enforcers. Anyway, Bell's men got hold of him and squeezed him to confess to blowing up a train."

"That is not small potatoes. A train? Even so, there is a lot dynamite that floats around this camp. Any half baked weasel could find enough to blow up something. Why did he do it, does it say?"

"He says he acted under orders of union leaders."

"Huh. I wonder how long it took Bell to force that out of him."

"It gets worse, Ben. The kicker—he said that those union officers were acting under the direction, control, and specific orders from Haywood."

"I find that hard to believe. Sure, we expected him to use the authority we gave him. But not like that."

"Authority?"

"Remember, back when we first met, Haywood was trying to get members to agree that local union officers needn't get members' approval for actions? He made several trips and finally got agreement on it."

"Yeah, I remember now. Said the officers needed 'flexibility.' Some flexibility, if this is true."

Ben frowned. "Oh it is technically true. He never gave explicit orders to blow up the Vindicator or a train. But he nods and winks and makes 'suggestions.' If you could hear him in private, you'd know he pulls all the strings in the union. And he hates mine owners. Nothing happens there he doesn't know about. Maybe not details but he has general knowledge of all. He approves everything. It is too bad. I'm sad his actions keep destroying things and people, rather than building a positive organization."

Abby closed the paper. "This low intensity war will change things. People are sick of being afraid to go out or let their children play in front of their house. The public will not stand for such violence and disorder. I hope this revelation doesn't hurt overall efforts at labor organizing. I expect it will hurt the Federation here."

A few weeks later, Abby's prediction came to pass. Public opinion turned against the union. So the sponsored violence abated. The WFM went silent. It did not stop work or fade away, it just laid low, had its enforcers take it easy. Some Cripple Creekers might have been lulled. Some thought the union had disbanded or fallen apart. Time passed. Citizens again felt comfortable walking the streets. Miners went down the shaft without fear. Ore came out to be shipped to the mill. Late fall turned into winter.

LON LAID THE PAPER NEXT TO HIS MORNING COFFEE. IT WAS dead of winter, early February. He noted the date of the paper, the second. The year was marching, plodding really, through storms and drifts, at least so far. Opening the paper, he saw the headline: 'Governor Peabody withdraws Militia.'

Good, he thought. We Cripple Creekers can get back to business for real now. We can forget the union rowdies and the arrogant soldiers with their stares and bayonets. I'll bet Clarence Hamlin's registry of anti union men is growing even longer. The next thought slid effortlessly in: Come to think of it I need to talk with that man. I'll head down there. After breakfast and reading the whole paper, he did.

"MR. HAMLIN, GOOD OF YOU TO SEE ME ON SHORT NOTICE." Lon looked around at the office. "Is your full time office here with the Tutt & Penrose Realty operation?"

Hamlin smiled. "No, I have a desk here is all. My actual title is Secretary of the Cripple Creek Mine Owners Association. People just call it the MOA. My primary job is to keep a list of workers who have quit the WFM. Or have agreed not to join it in the first place. I issue them work permits. Most of the big mines require an applicant to have a permit. That way we avoid paying WFM men to spy or make trouble."

Lon smiled. "And this has opened the door for you to

slowly strangle Big Bill Haywood's Western Federation of Miners."

"Mostly it lets us keep tabs on the good workers. But I agree with you, choking the WFM is a beneficial side effect. But you didn't come to discuss nuts and bolts of the labor wars. What can I do for you today, Mr. Bosini?"

"Well, I own the Double I mine. Over on Strombo Point. We're up and moving enough ore to pay our bills. It is time for us to join the MOA. It is important that we mine owners stick together, help each other out. And do what we can to keep the union under control. After all, men trying to make their lives and jobs better is one thing. Men bombing and killing quite another. I want to make whatever small contribution I can to that. And I plan to hire more workers soon, so I want to participate in the permit system."

"Welcome! Here is an application form. The dues vary on the amount of ore moved, so I can't tell you how much to pay yet. When you're ready to put men on, I will work with you to be sure you hire only men who are on the list. All are non WFM or other union."

A WEEK LATER, IN THE EVENING, ABBY KNOCKED ON LON'S door.

"Abby! So good to see you. It has been a long time. I feel bad…"

She barged in.

"What are you doing, joining the Double I with the MOA? Why do we want to do that? Why didn't you ask, discuss it? Mister big shot, I own half of that mine, don't you forget. And I want nothing to do with those capitalist snakes! Get us out."

She turned to go.

"No, recall our conversation. If you won't or don't do that, look at the papers.

You'll see that I own 50.1% of the stock. You know that. You agreed. We talked and since I found the vein and I work it every day I make the decisions. The Double I stays in the MOA."

The hostile glares were almost enough to give a sunburn as they stared at the other.

Lon broke eye contact. "Abby, can we talk?"

She looked away but gave a swift nod.

"Abigail. I know you think the world of Ben. He seems— is—a good man. But that union..."

"I don't like the violence either, Lon. But I am with Ben, wherever he goes and whatever he does."

"I hope that is not a mistake, Abby. Lots of folks will get hurt. But you are my sister. I accept your choices and want to be able to talk with you and not fight. And I want to be friends with Ben, at least I don't want to be his enemy."

"Don't you understand that your being in the MOA means you won't, can't hire Ben or his friends? Is that any way 'not to be his enemy'?"

The hot staring resumed.

"He runs with killers and thugs. Is that the kind of person you want spend your life with? To become?"

"He doesn't run with killers and thugs. There are some around the union, yes. For that matter, Dillon Bosini, there are killers and thugs hanging around the MOA. Is that the kind of person you want to be around, become?"

More hot tempered staring. Lon smiled, unamused.

"Touche, sister. It isn't simple, is it?"

"As I have been trying to tell you, Ben is actually a restraining influence on Shuler and Haywood. His influence somewhat balances out, offsets the crazies who want to shoot the place up. He gives advice. And they often take it, believe it or not. He always wants and says to go slow, don't lash out, think twice. If he weren't there, things would be worse."

"I just don't want you to get hurt." Their eyes were still locked, but the glare was gone.

"I'll tell you what, little sister. Sorry, old habits die hard and all. Tell you what, Abby, how about this: I'll hold on to the application to the MOA for now. I won't put the Double I in. One condition. If things get worse, if Haywood takes the union over the line again, blows something up or uses fists, all bets are off. I will then put the application in and do everything I can to protect you, me, and the Double I. And Ben and other workers. Alright?"

She halfheartedly jabbed, "Don't call me little sister." But she smiled as she said it.

"Alright, Lon. Maybe the three of us can gather for dinner?"

"Yes, let's try to find time."

X

WINTER TWO MILES ABOVE THE SEA IN THE SHADOW OF PIKES Peak is long as an Amazon python. Every year it squeezes the miners and club girls and owners and merchants as they face day after cold windy day. When people can barely take yet one more, warm and longer days come. The python loosens its grip.

Ben rode the lift up and was glad to feel the bump as it came to a stop. He stepped out, enjoying the springtime warmth. It was May and even at ten thousand feet, buds were swelling on the trees. He thought of Abby's flower garden, sporting a few daffodils and crocus. Color was nice to see after so many snow drifts and frigid nights.

He took his time walking to his rooms, savoring the day. "McNall, hello."

It was the new guy at the house. He wasn't all that new anymore; had been around for months. Name of George Henderson. He and Ben had kind of gotten acquainted over meals and occasional evening talks. His family, parents and siblings, lived over in Colorado Springs.

The man had a story, and Ben always thought of it when they ran across one another. Seems in the 1880's his father was Sheriff in Saguache County, about seventy five miles west of Cripple. The county was almost as big as Rhode Island. That was before the railroad came to the area, and Sheriff Henderson rode his county on horseback. He was in on the scandal of the decade. He was Deputy Sheriff at the time Alferd Packer's malevolent acts came to light. Packer was a would-be miner.

One spring day the man wandered into town, fat, happy, and alone. Henderson and others learned that he and several fellow miners had set out from Lake City, to the west, some months before. It was thought that all were lost in the winter storms. But no. A fascinating but repelling story unfolded: the man had killed and eaten his fellow travelers when caught in a blizzard. Later, a judge accused Packer of eating almost all the Democrats in the county. Packer was in and out of prison and had several trials over the affair. He was still behind bars, far as anyone knew. The whole tale again ran through Ben's mind as he saw George approach. He grinned in spite of himself.

"What's so funny, Ben?

"Just seeing you brought to mind the story of your dad and Packer and all. For some reason it strikes me funny. It isn't, I know. Better to laugh than cry, right? Anyway, hello, George! I haven't seen you around the house lately. Have you moved? Are things going alright for you?"

"No I haven't moved. I'm just on the go a lot. And yes, things are going well for me, thanks. I have hired on at the Findley Mine, over near Victor and Goldfield. It is a haul to get there I have to say. But I like living in Cripple and don't want to move to that side of the district. Likely would have trouble finding as nice a place there as I have here. Maybe couldn't, don't know. I like it over here. Plus, it is easy enough to ride the train to work. It is simple and cheap and safe. As you know, our rooming house gives us a good deal, decent plentiful food and clean. And the boarders are all civil and sane. Something to be said for that!"

Ben eyed him. "So you hired on at the Findley. I was there for a while. Had to leave, wouldn't do what they wanted to give me a permit. You must have gotten yourself on Hamlin's list."

"Yup." This was thin ice, a subject people were loathe to bring up. When it came up folks tiptoed around it, not to wake a sleeping mastiff. "I'm fairly new in town. Needed a job. Couldn't see swimming upstream against the powers that be."

"I can see your point. Was that hard to do—get on the list?"

"No. Just had to write a short sentence, not hard at all. Only thing I had to do besides give my name was the union question. Had to certify I wasn't WFM or any other mine union. That's the only way a man can get hired there."

"So how is it? Sometimes you hear the nonunion shops are hellish."

"Not at all, Ben. At least not the Findley I work at. There are no devils poking us with pitchforks. For me it is a good place. Decent pay with good crewmates and experienced, lenient foremen. Well, not lenient. That is too soft. But they don't treat us like we're jailbirds and they're the guards, if you know what I mean. Everyone I know is sane and knowledgeable, the type of man you want with you down the mine."

Ben had heard this from others. Most mines, he was told, were alright places to work, union or not. Hearing that always made him pause and wonder about his WFM loyalties.

"Ah. So the place is run by someone with his head screwed on straight?"

Henderson nodded. "Yes, that's a good way to put it. One thing, though, the Findley works round the clock. So we all have to take the night and day shifts in rotation. I don't care for that. Especially going in to work middle of the night. There's something, I don't know, sad about standing around after midnight waiting for a train to the mine. Especially the whistle when it stops to pick us up."

He shook his head, marveling and embarrassed that he had shared feelings and emotions with a man he barely knew. Feeling a little awkward, he added, "But it is alright. Work is work you know."

"So true. Well, good to see you, Henderson. Glad you are keeping busy and happy. We'll see you at the house when you have the day shift and come to regular meals."

He smiled. "Thanks. See you around."

Talk of the Findley and how Henderson liked it made him think. He had to sort things out, so he did what he always did then. He detoured from going home to see Abby. Talking with her always helped him put things in perspective.

"I SAW A MAN FROM THE ROOMING HOUSE ON THE WAY HOME. He is a scab at the Findley. Hamlin rules the place—you need to be on his list before they will hire you there. It is too bad. He and I could be friends I expect if he weren't working there. He matter of factly told me he had to disavow the WFM or other unions before he could get the job."

"It is hard to think about that, isn't it?"

"Yup. But he is like so many others. We are all just men trying to make a living. Too bad, he seems like a nice guy."

He thought a few moments. "I hope no more crazy men get their hands on trouble. That'd be bad. Boy, Abby, I sure hope the lid stays on. If it does, sometime down the road we can get the owners to talk and be reasonable."

"You didn't come here to tell me that."

"I came to talk to the prettiest woman in town."

She rolled her eyes but smiled.

"Sure you did. And I'm Queen Victoria. Let's sit and have coffee. And talk."

He sat, enjoying the drink and the company, and sorted his thoughts.

"Well, things are complicated. As you know, Haywood called a strike of the workers at the mill in Colorado City. The mills, really, there are more than one, and they process almost all the ore from the big boys here in Cripple."

"Not the miners here? What is the strategy?"

"He figured if the ore couldn't be processed the owners would feel the pinch. He wants to kill, or I should say make a hostage of the golden goose so to speak. The thinking is, they will come around and talk on our issues if they aren't making money. They will talk serious on a pay increase, pay for all hours worked, and so on. He wants to hit 'em where they hurt, you know."

He sipped and put his cup down. "One basic of union labor is mutual support. If one group goes on strike others do all they can to back their fellow workers. So he called on the miners here in Cripple and Victor to go out on sympathy strike. And other camps as well, but he is concentrating on Cripple."

"Yes, I knew that."

Abby stood and paced.

"And I know that quite a few miners refuse to strike. Many here, maybe most, are happy with their pay and conditions. They're unwilling to go out to support twenty or thirty workers they don't know and never see. Many miners here have joined Hamlin's list, like the man you saw a while ago. And are going to work."

Stopped, she looked out the window at Pikes Peak. "The union calls them strike breakers, or worse, scabs. But they just call themselves miners."

She sat and swigged the last of her coffee. Then she looked him straight on. He had a flash of a doctor preparing to deliver bad news.

"Maybe Big Bill has overstepped himself this time."

"That may be, Abby, that may well be. I know there is a lot of animosity, fury, hatred going around. The WFM and the MOA are locked in a hateful embrace. They need each other but neither can let go long enough to shake the other's hand and start doing some positive work. It really is too bad we have to choose sides."

"I agree, Ben. Yes there is a lot of hostility between the union and non union points of view. And it is not getting any better. No one knows where this will all end up. But I want you to know, Ben McNall, that I am with you. Whatever you decide, I am with you."

He stopped thinking about labor and management and concentrated on the person with him. Her ocean deep eyes almost swallowed him.

"Thank you for that, my love. I hoped to hear you say that. And I hope we are always with each other."

The look lasted a moment longer then he looked away.

"I feel better about things. There will always be union and management, up and down, right and left. There will always be something for people to fight over. But no matter. Our lives will go on. It is always good to talk with you. Hey, I need to go home and clean up. Can I come back later?"

"You had better, mister. See you in an hour or so."

BEN COASTED TOWARDS HOME. HE WAS IMMERSED IN A HAZE of love and warmth. He was thinking, feeling that boy oh boy did he love to be with Abby and talk with her. She just plain made him feel good. He floated on down the road on a cloud, lost in thoughts of brown eyes and red hair and an extraordinary woman.

Even though the sun was out, a shadowy shape suddenly came from nowhere and walked at his side.

"McNall. Ben. Hello."

Owens Anderson was by nature fidgety, expressive and full of movement. Not now. This day he was positively subdued. Ben wondered if he was in shock or something. Without moving his head he looked around constantly, almost guiltily. Hands in pockets, he all but scrunched down as if to make himself less visible.

"My man Owe! Hey, it is good to see you. Haven't talked with you since our mysterious meeting last fall. When you gave me hints...Say, I've wondered and wondered, how did you know about the Vindicator? Beforehand? That's what you were warning me about, wasn't it?"

Owens grabbed his elbow and hustled him into a side street. "Shhh! Don't even say that word, you'll get us in big trouble."

He shook free and stepped back. "Oh? 'Us?' Why? Should I be worried about you, being seen with you? What have you been doing, my friend?"

Suddenly the old Owens, the one he knew, came out of the sullen and frightened shell. He smiled, nodded, waved his arms, and strode in small circles. It brought to mind a man happily trying to take gloves off while walking with one shoe nailed down. Ben watched a moment then choked a giggle.

Owens started to talk, the old loquacious Owens.

"My old friend Ben, who I came to town with years ago. It's like this, my man, I won't tell you what I have been doing 'cause what you don't know can't hurt you so if anyone asks, you tell how you and I came to town several years ago and went our separate ways and you don't know what I'm doing or where I am, I'm serious, do you got that? You haven't seen me lately and have no idea what I'm up to, you got that?"

Deadpan, Ben asked, "Excuse me, sir, who did you say you were?"

"Hah. Seriously." The smile faded. "You didn't hear this from me. Like I say, you haven't seen me in ages. But if someone asks and you have to say, just say you heard it somewhere, not sure, you know, people blab and all. Anyway."

Owens looked around again, a rabbit checking for coyotes.

"Ben. There is talk of another kaboom."

"Kaboom? You mean another explosion?"

Owe looked his friend full in the eye and slowly nodded.

"Not in a mine. On the way to one, out in the open. You know now that my friend won't hesitate to blow things up and he hates scabs and he intends to finish off a bunch of them and if some others are hurt, well that's just too bad 'cause he wants to send a message and send it big."

"But if you know about this, why don't you stop him? Tell someone. How can any good come of such a thing?"

"Ben. That is why I'm talking to you."

He paused, still for a moment.

"I can't stop him myself and he says the owners are using men like rag dolls and if anything needs stopped it is that and we have to draw attention. Plus, he is my friend and I have to do what he says and if anyone learns I told you this I am a dead man and, and I've said too much already I gotta go. Take care, Ben."

Owens became a shadow again; the wraith disappeared.

Nonplussed, Ben stood a moment. Then he returned to the main street and sat on a bench, pondering. After a few minutes, he decided, stood and ambled to the union offices.

He nodded at the secretary/receptionist. "I need to see Shuler. Is he here?"

"No, all the officers are at the annual convention in Denver. It is held there every May, usually the third week. They will be back later. Frank I think will return tomorrow, some of the others later in the week. Can I take a message? If it is urgent we can telegraph or even make a telephone call."

"Nah, I just had some scuttlebutt I wanted to share. I'll get back to him. Thanks."

HE BACKTRACKED TO ABBY'S AGAIN. HE LOOKED GRIM; SHE smiled then silently waited for him to say something.

"Another warning, Ab."

"Oh no Ben. What?"

"One of the same guys who hinted at the Vindicator came up. When I said the mine's name he told he to keep it quiet or he'd be a dead man. So he is in deep with something."

"Yes, it sounds like he is. I hope he doesn't pull you in too." Ben ignored that.

"Anyway, he said there will be another event. An explosion, 'kaboom' as he put it. Not in a mine this time, out in the open he said. Whatever or wherever that may be. To get some scabs, he said."

"We have to warn Lon."

"No not yet. That is too general—an explosion out in the open to kill workers. Hell, we all know there are madmen who want to do that. And how do you guard against that? I don't want to be the boy that cried wolf, you know? Let me try to get something more specific. Let me listen and watch and dig. Give me a few days."

She didn't like it. "This is a credible source, Ben. I think we ought to act now, but alright. Not a few days, though. I'll give you a day. One day."

He didn't like it but couldn't find a good argument. She took his silence for agreement.

"Alright, I see you agree. Now, hear me. The day after tomorrow, first thing, I'm going to Lon and tell him about this. You have until then. And be careful Ben. If your friend is in danger you could be too. If you get in the way, no telling what someone might try to do to you."

He gave in. "Fair enough. I have work to do, have to go talk to someone. See you later. Love you."

He was out the door before he finished this.

Across town Juni was enjoying a nice day. He was inside, doing something he loved: shooting billiards. Pool. It was his relaxer, his salvation. His friends, people he enjoyed and relied on, all played it. It was recreation and one of his favorites.

He admitted to having another favorite, what he did to make a living. The popular name was high grading. He did not call it that. He was an independent small business man, an ore mover. And it made him good money. It gave him a flexible schedule and days free. For pool.

He spent his sunlight hours (not that he cared to actually be outdoors in it) at one particular saloon with three fine, level slate tables with untorn felt covers. The owner didn't let stray miners sleep on them and there were no girls for distraction. Aside from drinking, the place was known only for billiards. The tables were ideal. Most days he was there with friends, racking and yakking. Good times!

"Three ball in the corner." He pointed his cue then lined up and shot, aiming to drop the three ball elegantly after caroming off of several cushions.

"Juni. Someone to see you."

Placid to a fault, Juni didn't upset easily. This, an interruption in the middle of a shot, troubled him.

"Hell." He waited for the ball to drop, which it did. He threw the cue down in anger but was careful not to rip the felt. The big man turned, ready to snarl. To the surprise of those around him, he paused then smiled.

"Ben! This is a surprise. How did you find me? Good to see you, it has been a while! What's up? Are you here for that lesson I offered a while back?" Juni grabbed his friend's hand and shook it vigorously.

"No, no lesson today, Juni. You are way better than I ever could be anyway. Thing is, I wanted to see you, and then I remembered that you brought me here. That time when you were trying to school me on breaking and carom shots as I recall. The lessons didn't take—I never was much of a student!"

He was able to pull his hand from Juni's grasp.

"What's up, you ask? Can we walk? This hall sure is crowded and noisy."

Juni glanced at his opponent. "I'll give you this game. Rematch later." He retrieved his cue and put it in the rack reserved for regular players.

"Let's go, this door takes us right out."

Ben cut to the chase. "I saw Owens a while ago."

"Oh? I haven't seen him in ages. How is Old Twitchy? What is he doing nowadays?"

"Old Twitchy? Where the...never mind. I'm not sure what he's doing, Juni. But I don't think it is good, whatever it is. Earlier today he appeared at my side like a ghost. In the street. Acted odd."

"Nothing new about that. The odd, I mean. Usually he is talking loud, waving arms and pacing around like a, well, you know what I mean."

"True. Anyway, he acted subdued, almost like he was being hunted. Very unlike him. And then of a sudden he opened up, and all but said there'd be another explosion. It really reminded me of his message just before the Vindicator killings. And yours too for that matter. But this will be bigger, he hinted."

"Oh?" The big man stopped, faced Ben.

"Juni, this is crazy. A big bad action could kill and hurt innocent people. And it will fracture the community. Half or more of the town dislikes the union and disrespects union men. And those feelings are mutual. If some madman does something awful the place could go to war. Do you know anything? Have you heard any rumors, hints of action?"

"Rumors fly all the time, Ben. That's all they are, is people talking and repeating. Things get expanded and guessed at and so on. Yah, and pretty soon we're scaring ourselves to death."

"You didn't answer my question."

Juni looked at him a moment, then reached into a pocket and brought out a piece of paper.

"Got a pencil?"

With Ben's pencil he wrote something on the paper, folded it once and dropped it.

"Here's your pencil." He looked down. "Oh look, a piece of paper. Could be a note of some kind. On the street. I wonder what it says. A curious man like you would pick it up and read it, I bet. Say, I have to get back to my game. Take care, Ben, see you around."

He spun, left, and the meeting broke up as fast as a nine ball set when hit by the cue ball.

"Juni!"

The man ignored him, striding fast, well on the way to his rematch.

Ben shrugged, pocketed the note and headed down the street, towards his room.

'IF YOU LIVE BY INDEPENDENCE LEAVE YOUR PLACE ON THE FIFTH'

Ben again tried to decipher Juni's scrawled, unsigned note. He'd been over it twenty times at least, turning the words this way and that.

Independence? Did that mean a frame of mind, the holiday, the mine, the town, its train station, a scab, a miner relying on himself, or what? And how did he mean, 'by'?

Leave your place? What place—home, job? Go where? For how long? Why?

Fifth—day of the week, anniversary of something, work shift, hour of daylight...?

It made no sense no matter how often he looked at it. After a while, he reached a decision, made a plan. First he copied the note word for word, a number of times so he had several to show around. The original he stuck in a book for safekeeping. He didn't want to inadvertently give any clue to the note's origin. The copies he folded and put in a pocket to take to work the next day.

Still, the wording haunted him. He had dreams of Thomas Jefferson standing in the plaza of a Missouri town holding the Declaration of Independence, yelling at people streaming down a street in groups of five, hurrying somewhere, some placid and some fearful, while lightning struck all around. It was not a restful night.

THE PLEASANT SPRING WEATHER CONTINUED. NEXT AFTERnoon was nice. Ben enjoyed the sun. Its warmth was a nice counterpoint to the gloom and dread he felt. He got off the lift and headed for the union office. He didn't know what was to be learned there. And he wasn't sure he was ready to hear it.

"I need to see Shuler—is he back in town?"

The receptionist nodded. "You know where his office is. Go on back."

Ben knocked on the door jamb and entered.

"Hello Frank. How was the convention?"

"Unh. Lot of anger and arguments. Even for a bunch of ornery union men it was more hostile and thorny than usual. Miners from all over the western states, bitching and moaning. Come on in and close the door."

"Lot of angry folks in Denver, huh?"

"You didn't hear it here, alright?"

Ben held a hand to an ear. "Hear what?"

Shuler smiled. "Most of the anger was directed at Haywood. He has spent a ton of money on this strike. And he is losing it. We are losing it. We heard it over and over: The mills are moving ore. Hamlin's list is growing. Owners in other camps are imitating his methods. People here in the district are losing sympathy for the WFM. And that is true in other mining camps too—public opinion is turning against us. Members want a settlement, something face saving to get us back to an even keel. But Big Bill, being the fighter he is, says no. Says we should double down. We are on the right side, he says. Maybe we are, maybe we aren't. What I'm afraid is, if we do win, it will be a pyrrhic victory."

"Pyrrhic?"

"You know, some historic Greek battle. Back before Jesus Christ. The winning side spent so much blood and money to win, it wasn't worth the cost. They came out too weak to enjoy or take advantage of their triumph."

"Oh."

Shuler went on. "This is not good, not good at all. I'm not sure where things go from here."

"I wanted to ask about that. Look."

Shuler took and scanned the note. He blanched.

"Where did you get this?"

"I can't really tell you. Suffice to say, it appeared at my feet one day. Peculiar deal."

Shuler reread the note, looking more shaken.

Ben waited a moment but got no response. "I'm not sure what this means, Frank. It could be a warning or alert, don't you think?"

"I suppose. Or it could be a child practicing penmanship."

"Oh?"

The union official gazed out the window and visibly made a decision. He looked Ben in the eye. "Ben. Sometimes the union has to look the other way as someone takes unpleasant measures. To draw attention and to protect the members."

"What are you telling me, Frank?"

"I'm telling you that even if things look dark right now, the struggle isn't over. There will be other actions I can guarantee you. This may be a warning of one of them. I don't know." When he said this, Shuler looked everywhere but at Ben as he spoke.

A pause. Then he said a strange. "Accidents happen, even at train stations, you know?"

He sat, still avoiding Ben's eyes, and fiddled with stacks of paper on his desk. "Well, I was out for quite a while. Need to catch up. Thanks for stopping by, McNall."

Ben felt like Alice in a hall of mirrors. He had to get out to sunlight and think. Nodding a 'so long,' he left. He needed to talk this out so he headed to Abby's.

"INDEPENDENCE MUST REFER TO THE STATION." THIS suddenly came to him as he looked between Abby and the note.

"Let me get this straight. You found this on the street and the union guy said it was just a kid. Come on, Ben. Any scrap you found on the street full of gibberish, you'd throw it down. Where'd you get it?"

"Keep a secret?"

With a nod. "You know that, Ben."

"I know, but had to ask. This could land you and me in big trouble if we aren't careful."

"Or if we are careful. What's the secret?"

"Well, after the one warning I told you about I went and found a friend of that source. I told him about it and he clammed up. Acted kind of strange and distant when I pressed him. After a minute he asked for a pencil, wrote the note out, and dropped it. Just dropped it at my feet."

Ben thought back to get the words right. "Then he said 'look, a note on the street. Wonder what it says? A curious guy would read it.' Before I could say anything or pick it up, he turned and left in a hurry. He ignored my calls and wouldn't say another word."

"So this is a warning."

"Yes."

They both looked again at the little piece of paper, almost as if hoping it would explain things. It didn't but Ben did.

"And from what Shuler said, I think 'independence' refers to the station, the train station at Independence."

"So something is going to happen on the fifth of something and if you are around you want to go away. Does that make sense?"

"Could be. This is iffy and nebulous as hell. I want to poke around a little more, try to learn what it means. We need to be sure before we decide to share this. Or less unsure at least."

Her eyes grew wide. "Ben. This is May 31. What if 'fifth' means June 5?"

"That means not much time. Makes sense, get out of the area on June 5. Maybe that's it. But why?"

"Whatever is in the works, it is not good if people nearby are being warned away."

"Maybe it is time to tell Lon. We need someone who can look with fresh eyes."

"Yes, let's, tomorrow morning."

They puzzled and talked more. At the same time, another meeting was coming together across town, near the very area they puzzled about.

OWENS WASN'T SURE HE ALWAYS LIKED THE WAY HIS FRIEND Harry acted. Matter of fact, he was sure of it. The man seemed quiet and confident and he sure knew how to rig up a blast. But he could be short tempered, sometimes even downright mean. And he didn't always tell the truth or give the full picture of what he wanted Owens to do.

Orchard had called him, told him to meet up at a certain time and place. So here they were, in an old building near the Independence train station. What was going on now?

"Alright, Anderson, good to see you. Glad we got the time right. Thanks for meeting me like this. Here's the thing, Owens. I will be going camping and fishing with friends the next few days."

"Sounds like fun. But, Harry, why are you telling me, am I going too or is there something else you need for me to do?"

"Well, no, Owens. You aren't going on this trip. Maybe another time. Funny you should ask about something else. Because I need your help with a project first."

"Alright, Harry, I can do that, what is it that we are doing?"

"You keep this to yourself now, you hear? This is a big project which will be famous but we can't tell anyone about our part in it. Never, not now. Not later. Not ever. Do you understand that, Owens?"

As he said this, his eyes were bright but the pupils were tiny black points like snake eyes, and they followed him as a snake ready to strike. Owens was skewered by their intensity, and felt a twinge of fear.

"Sure Harry, I'm in and my lips are riveted shut."

Orchard nodded and his eyes relaxed. A little bit.

Owens still wondered. "What is it?"

"Tonight after dark meet me here. Say at ten o'clock. We will put some goods under the waiting platform over there." He nodded at the station. "And we'll run some wires over to here. We'll do that when it is too dark to be seen. So wear something dark that will blend in. And not a word to anyone, right?"

"Sure, Harry. Ten, here, got it, see you then."

XI

Abby lay in bed pondering the note. Again. It could mean most anything, she thought. What if it really was some kind of a penmanship exercise? No, that wasn't a child's handwriting. And Ben wouldn't be upset if it were. There was something to it all. With half a shrug, she nodded off to sleep.

Dreams came. She was on the fifth floor of some building, some strange structure she had never seen before. She was among a bunch of people, some she knew and many, not. They had to leave, she had to leave, no one knew why but they had to get out of there, and there was a crush of people at the one tiny window which would let them out there was no door and she couldn't breathe for the closeness and someone or something let out a high shrill yell. A stampede to the window started, worse than before. But the shrill noise kind of sounded like a cat, maybe a mountain lion, and there was thunder or drums beating or something.

Insistent knocking at the door finally woke her. Her eyes opened only to see big cat eyes staring. Her pet sat square on her chest, not two inches away, intently glaring as it meowed again. Abby shoved the cat off and blearily sat up. It was Sunday. June 5. What a dream!

Slipping on a robe, she went to the door and listened. Quiet. She was sure she heard someone knocking. Swinging it open a crack, she belatedly hoped the knocker was either Lon, Ben, or one of her teacher colleagues.

"Ben. What are you doing out at dawn?" She propped the door against her foot and drew her robe tighter. A glance to the street told her no one saw them.

"Come in." She stepped back and really looked at him.

"You look as groggy as I feel! Did you sleep at all? I did but boy oh boy the dreams were weird and scary."

Just awakened, it took a moment for the dream to sink in. "And somehow the dreams were tied to that darn note. I need coffee, do you want a cup?"

"Uh, yeah, please." He smiled. "Thought you'd never ask! Yeah, I slept. But like you, not real well. I too had a dream. Several different ones. The worst was, a big explosion. I was in a train car I think, or maybe a stage coach, not really sure. Couldn't tell if we were moving somewhere, on tracks or in a mine or what. And Lon was on board with me—not sure where you were, somewhere around I think. There was a big bang and confusion and he, well, after a while he was gone."

Each regarded the other over their coffee. Neither said it, but the other looked as if they had lived rather than visualized their dreams. Both could have gone back to bed for a few hours.

"Oh Ben. We agreed to tell Lon. We really need to, soon. He needs to know about all this. I know you want to figure out the note. So do I, but this is bigger than you and me. It is eating us up. We need some other folks to look at it and work the problem. At least he can put the word out and maybe if some monster tries something he can be stopped.

He'll be here at about nine to take me to church. Let's talk to him then."

He stood and paced.

"Yes, you're right. Let's. We need more ideas, someone to take a fresh look. I see no point in losing our minds over this."

"Tell you what. I'll go to my room and get dressed. You sit back down and have another cup of coffee."

He reached for her hands and helped her stand up. "I have a better idea Abby. Let's the two of us go to your room."

She smiled and leaned in for a kiss, then led him away.

ABOUT TWO HOURS LATER THEY AGAIN SIPPED COFFEE, INVIGO-rated yet relaxed as only lovers can be. An exquisite summer morning, warm and sunny, topped off the mood. The expected knock was not long coming. Abby looked to Ben and he nodded as he stood to answer it.

Lon was surprised, embarrassed, shocked when it registered who swung the door open. He could barely form a question. Later they would laugh at how lame he sounded.

"Uh...Hello Ben." A several second pause. "What are you doing here?"

He looked helplessly at his sister, anger starting to grow.

"He's here because we have a problem, Lon. I asked him here." Her steady gaze dared him to speak and his temper cooled. He said nothing more, simply walked in and looked between them.

"Lon. There's another warning. A credible one, it seems. We've been tearing our hair out trying to decipher it.

Turning it every way but Sunday, having nightmares even.
We need help."

She repeated herself. "He's here because I asked him
here, Lon."

He held hands up in surrender, looking between them.
"I didn't say anything. Was surprised is all, can't blame me
for that. This is between you two."

He took a deep breath, smiled half heartedly, and asked,
"Now, what is this warning?"

Ben handed him a piece of paper then sat down. "Read
this. Maybe it will make sense to you. Don't ask how I got it.
All I can say is, it came from a credible reliable source."

He read and reread it, thought a moment, slouched into
an easy chair. "Hey, how about some coffee, Abby?"

Her first inclination was to tell him to get it himself, then
she really wanted it pour it over his head. She iced the anger
and poured him a cup. After all, she did ask for his help. He
was handling Ben's being there pretty well, and thankfully
there were no shouts or flying fists.

"So something big and showy, probably bad, is happening, maybe in Independence, on or around the fifth. Of
something." Lon sat up straight. "That could mean today!"

"Bingo. We think it will be at the Independence station.
Probably an attack or show of force of some kind aimed at
strike breakers." Ben shifted uneasily in his seat, looking and
listening to Lon's response.

"Strike breakers hell. They are miners who choose not to
associate with the WFM." He glanced at Ben who shrugged
and commented.

"Whatever we call them, they are people who shouldn't
be harmed."

Lon sighed. "Well, maybe, that narrows it down. Innocent people, maybe in the crosshairs. What do we do to stop that?"

"It helps to know what you are trying to stop." Sorrow and frustration oozed in Ben's voice.

"Maybe we don't stop it, whatever 'it' turns out to be. I say we don't even try to intervene."

It was Abby's turn to stand and pace. "Maybe we just warn miners to be extra alert today and tomorrow. If anyone can take care of themselves, it is a bunch of ornery miners. God help the fiend who tries something. A sane person would think twice about taking on a watchful angry miner, much less a bunch of them."

Ben: "Sane. That is the thing, is the person or people behind this playing with a full deck?"

Lon rubbed his face, suddenly feeling tired as Sisyphus. "Yeah. I agree with you Abby. Don't try to stop whatever it is. We don't know or understand it and maybe it is nothing. Just get the word out that something may happen. This is our best bet, probably the only bet. I'll try to get the word out to owners and the foremen at all the mines. Ben, maybe you want to talk to your friends and contacts too. Don't endanger yourself but the more who know about this, the better."

Ben nodded. "Makes sense. I will tell some folks. It'll be bad for me if I'm not careful and everyone will be after my hide. If things go the wrong way, I could end up in the crosshairs of both the WFM and the MOA. Not to mention maybe a mad man out there with dynamite. Wonderful."

Abby blanched at that. Lon didn't hear. He was still thinking about a warning.

"I'd love to get a notice in the papers. It may be too late for that but I'll try."

Before this trio met, after night fell, events took on their own momentum. Even before Abby's cat snuggled and Ben dreamt, men were busy on another side of the district.

Near the train station, two shadows converged. One man pulled out a watch and peered closely. Owens Anderson had a stray, for him profound thought. The time checker could have been a merchant checking on the arrival of his chief clerk. All in a day's work and so forth, nothing unusual happening. The man spoke softly. "You are right on time, Owens. Good work."

"Well, that's when you told me to meet you here, Harry."

"Shhh. Not so loud, my friend. Quiet is the thing! Now, we have to get to work." He pointed. "Take one of those cases, careful like. It is full of dynamite. Don't drop it!"

"Gee, really Harry you mean I can't treat it like a box of laundry soap?"

Orchard glared. "Just take one. I'll take the other. We're going over there to the Independence Station. Its where the scabs gather before they get on the train for their shift. It is only fifty or sixty yards, so follow me and watch your footing. If you have to say something, don't say nothin', just tap me on the shoulder so we can whisper."

Quietly and carefully they went to a spot under the platform at the station. With careful pushing and shoving, the boxes were positioned just so.

Harry whispered, close to Owens' ear. "I'll wire this up. You unspool the wire, lay it on the ground so it doesn't catch light or shine, and run it back to where we were. Quiet like, no kinks in the wire, and be sure it doesn't get cut or broken. I don't need to tell you not to show yourself more than you have to."

Orchard connected his end to the explosives. Owens unspooled the wire and was steadily moving back where they would wait. Harry attached the ends to a wooden crate, rigged to tilt when the wire was pulled. That in turn would pour acid onto the dynamite, setting off an explosion. The man regarded his arrangement. He had to admit, it was elegant. And there was a whole lot of dynamite, two boxes full. The set up was efficient and reliable and had been used before.

Harry loved that there was a lot of kaboom smack dab in front of him. He eased back, admiring his work. With a smile he started back, feeling, keeping a hand on the wire to be sure it wasn't snagged or somehow broken or torn. It wasn't that he didn't trust Owens to do the job. Just wanted to be sure. Plus, he loved the feel and anticipation of it.

"Alright, now, Owe. Things are about set. Let me fix a stick leg to the wires to give us a handle to pull on. When we are ready to close the circuit."

Owens watched Harry Orchard, the mystery man on the run from Idaho, high grader and explosives expert. The man deftly wrapped the end of the wires around a curved wooden rod probably two feet long. It was a chair leg he brought and left there just for this purpose. He made sure to get it snugly attached but was careful not to tug or put pressure on the wires.

"Alright, Owens. We're done for now. Ya done good! Now, you go back home quiet like and get some sleep. Meet me here, same time tonight, about 23 hours from now, midnight. And not a word of this to anyone, like I've said."

218 | STAN MOORE

"I thought you were going fishing?"

"Oh, I will arrange for a horse and wagon. I'll be sure to be seen with it. You probably don't need to know, but later this morning, in a few hours, I will leave town with a friend and his son. The wagon will be loaded to camp and fish and all. But don't worry. I will come back and meet you here at midnight. Oh, one other thing I need you to do: tonight, bring a small pail or bottle of kerosene and some rags to tie around our shoes. Very important, can you do that? See you tonight, Owens."

"Yes, I'll be here. Harry? Why are we doing this? What are we going to blow up?"

Orchard's face hardened, focused with rage and hate.

"We're doing this because Bill Haywood's strike has hit a slow patch. We need to knock things off center, off balance, stir things up. And kill scabs."

"Why focus on scabs? Aren't they just men trying to work for a living?"

"No. Scabs are a poison in the system. They are the cause and the symptom of our problem. Any man who would cross a picket line, or turn his back on his union brothers, is scum. And he needs dealt with."

"Oh."

"Plus, Owe, I have to tell you. I have always liked explosives. I love to make a bing bang boom. Big or little, God help me, I love it." His smile faded.

"But that is just a side benefit here, a blast. Really, with this, we're hitting at the rich mine owners and their scabs."

"If you say so, Harry. See you tonight. And I'll bring kerosene and rags."

Lon left Ben and Abby sipping coffee. He was in a hurry. He needed help and his first stop was with someone well connected. If anyone could get word out, it was this man.

"Mr. Hamlin, I know you are a busy man. But can I have three minutes of your time?"

Hamlin was on the way out of the Tutt & Penrose Realty office. He stopped, giving Lon an appraising look.

"Important, is it, Bosini? Well, walk with me." Off he strode.

Lon caught up. "I think we may see another action, a WFM action. Today or tonight."

"Oh?"

Holding out the paper, Lon explained. "This is from a reliable but anonymous source."

"Anonymous? What good is that?"

"Let's just say the WFM plays rough and people have a life and want to keep it."

Hamlin read the note once and again then handed it back.

"Hmph. Any suggestions?"

"Put people on alert. Do our best to tell folks. Tell them to be aware that someone might try to hit at strikebreakers. Encourage them to tell the Sheriff if something is suspicious. I don't know what else can be done on this admittedly scant information."

"The Sheriff. That union loving, Federation sympathizing fool. Well, I guess no harm can come of it. Give the note back. I'll keep it and will pass the word to our members."

Lon nodded, stopped. Hamlin went on a step, stopped and turned.

"Bosini. Thank you." He strode off briskly, headed somewhere in a hurry.

––––––––––

BEN AND ABBY TOOK A WALK. EVEN WITH THE NOTE AND ALL, they wanted to take in the sunny spring day. Town was two miles above the sea and hemmed in by big mountains. They, the peaks, made their own weather which was often cold and stormy. A clear sunlit day like this was to be savored.

"Abby, this life is getting pretty intense. Maybe I should work a deal with Lon to buy into the Double I. Then we could marry and maybe move to Colorado Springs or Denver. Get away from this hostile, polarized, tense gold camp. Spend time doing something besides ferreting out strange and murky threats."

Abby stopped, looked him deep. "Yes."

"Yes? Yes what, Abby?"

Her laugh was a beautiful lilt. "Ben McNall, I figured you for a smart one. But sometimes I wonder if you could find your way out of a closet!" She laughed again.

"Yes I will marry you. If you want to leave the mines and become an owner, yes. If you want to stay in the mines, yes. If you want to become a saddle tramp and herd cattle in Montana, yes!"

She laughed again and grabbed his hands, which was a pretty daring display of affection in public.

"As soon as we can, we will." He smiled, dazed, and they walked on.

––––––––––

THE DAY WAS A PLEASURE TO ORCHARD AS WELL. THE SUN'S warmth was exquisite on his back as he approached the stables.

"How can I help you?"

"I need to rent a wagon, and a riding horse too." Orchard smiled at the blacksmith.

"How far are you headed with 'em?"

"Oh, a friend and I are taking his son out fishing and camping. Going out south of town for a night or two, then on to Canon City."

"That'll cost you three dollars a day, two for the wagon and team, one for the horse. Two day's deposit, and you can pay the balance when you turn 'em into our Canon City stable. You are responsible for feeding and watering, and any injuries."

"Done. Here's six dollars." After the money passed they shook hands.

The smith motioned. "There's the wagon. You want to ride or have the horse follow?"

"Follow. I'm going to buy supplies. Should be a good trip, I'm looking forward to it. We're headed to Millsap Creek, or maybe Mill Creek out there."

The smith smiled. "I pulled six or eight nice trout out of Mill a few weeks ago. You will do well. Your friend's son should catch a few. There are lots of nice holes full of hungry fish! Good luck!"

Later that day and a few miles south, a heavily laden wagon stopped at a clearing. Orchard, his friend, and the man's son got off.

"This looks like a good place to camp. Let's pitch the tent."

Camp set, dinner done, it was a pleasant night. Orchard looked at the boy, smiled. "Want to play a game of mumblety peg?"

"Sure." The boy got out and opened his pocket knife. Balancing by the tip of the blade, he flipped it.

"Double flip and perfect stick. Top that, Mr. Orchard!"

The game went on like this for a while, the boy giving better than he got.

"It is getting late. You win. You really can handle that knife, boy!"

"Thanks. I can't wait to tell my friends I beat you!"

Camp settled down. The fire burned low, embers deep. The man and son turned in.

"I think I'll sit up and watch the stars a while. Good night."

"Night, Harry. See you in the morning."

Orchard sat and enjoyed the glow of the embers. The reds and purples swirled hypnotically. He sat quietly, listening for any movement. After he was sure the camp was asleep, he stood and went over to the horse. Dangling a carrot to keep her quiet and busy chewing, he saddled her. Another carrot then he untied and walked her away from camp. When he was out of hearing, he mounted, settled in, and spurred the steed back north.

IN THE EVENING, LON MADE THE ROUNDS OF THE CLUBS, AT least the more or less reputable ones. Every small mine owner he knew or knew of got the word. Also he spoke to the merchants and saloon keepers he met. He told in general terms of the threat, and urged them to alert their people. Most were skeptical but saw no harm in being watchful.

BEN AND ABBY WERE TOGETHER, HARDLY TALKING, OCCASION-
ally nuzzling, looking into the eyes and soul of the other.
They couldn't have gotten any closer. If either spoke, it was of
the future, their future. The evening went by slowly, quickly,
gloriously.

OWENS LOOKED AT HIS WATCH. HE WAS SURE IT WAS RIGHT,
and it said midnight. Where was Harry? He stewed and
worried and paced. Until Orchard appeared and tapped his
shoulder, Owens noticed nothing but his fret.

"Harry!"

"Shhh. Whisper, don't talk. We can't be seen tonight."

Anderson nodded, pointed. "Like you told me. Kerosene. Rags. Also some string."

Harry nodded, doused rags and tied one to each shoe.
Owens followed suit.

Orchard motioned and they carefully went to where the
chair leg was wrapped with wire.

"Now we wait, Owens. Not a word from you, alright?"

Owens nodded, uncomfortable at the thought of blowing
up something but excited too.

They sat on the ground. It wasn't comfortable but they
made the best of it.

Owens Anderson was usually on the move, talking, gesturing, nodding, hands fiddling. Even on a quiet, normal
night he would have had genuine trouble sitting mum and
still in the dark. Tonight was harder, exciting as the scenario

was. Not that he knew the word scenario, but he sure knew this was not a typical night in Cripple Creek. He tried to count railroad ties, and the stars, and the number of houses he could see. Many, probably most of the houses he could see had at least one light going. After counting and recounting all within his line of sight, he realized that the nearby houses were dark. He couldn't help himself, he had to speak. To his credit he leaned close and whispered.

"Harry, all of the houses around here are dark. Out of the whole camp, this is the only fully dark stretch. How odd. Almost like they're empty, like people know the sky is going to shatter."

Orchard gave him a 'you poor dumb bastard' look. "Hmm. What a coincidence."

A crowd was gathering on the platform. Workers at the shift change.

Orchard bristled. "Bunch of vermin, those slimy scabs. God damn then."

The train approached, light shining on the men who gathered at the platform to board.

"Get ready, Anderson. Cover up, this will pack a big punch." They shook hands. Orchard had one last loose end to tend to. He looked Owens in the eye: "You're going back to bed and I'm going fishing, right? And never a word to anyone!"

Both men laid flat. Owens covered his head with both arms. Orchard put one up over his head. With the other he firmly held the chair leg holding the wires. When the engineer blew the whistle signifying a stop at the station, he yanked.

XII

MINERS WAITING FOR THE SHIFT CHANGE KILLED TIME ANY way they could. Some thought of the work to be done, braces to be placed or holes to be drilled or a face just blown to be mucked. Others thought of the mining claims they worked during their off hours, hoping to hit it big and be done with shift work. Many, perhaps most, thought of home, family, the day's occurrences, and general woolgathering.

This miner's random thoughts while waiting through the shift change are typical. He could be one of hundreds in and around Cripple Creek. Or in almost any other gold camp.

TODAY AT THE FINDLEY MINE WAS BUSINESS AS USUAL. WE had the normal drilling, blasting, mucking, noise and darkness. Good money. And pretty good conditions with knowledgeable foremen. They were just miners themselves once and know how the mine works. That isn't always true of owners, but ours do know mines and mining. All in all, they don't wear horns. Satan does not own our mine. We have realistic production goals and they pay us pretty well.

Labor. It means work, not politics. Or should, anyway. If there are any union lovers around our place, they are keeping quiet. Most of us just want to work an honest job and to be able to take care of our families. We don't see the need to pay some union officer to tell us what our wage should be or how many hours we should work. We can figure that out for ourselves.

This morning's warning was unusual, odd. In a few words, all it was, really, was, 'something may happen.' There is so much back and forth, so much tough talk by the union and the owners. I wonder why today they warn of something. Must be something to it, or at least someone big thinks so. The foreman did his best not to laugh when he brought it up. I think he hid a grin when he told us. 'Be alert for someone trying to harm you today the fifth of June.'

No kidding! Let's see, we're in a camp with forty or fifty thousand people, mostly men. There is gold flowing like water, liquor flowing like gold, most everyone carries a gun, sleep is in short supply, mining claims are argued over, and the mine owners are fighting with the union men. Everything is running full tilt twenty four hours a day—liquor flows, mines and smelters are working, the dance halls are busy, stores are open, streets are full at all hours. Add to that, some people like the union, some hate it, and most wish the whole damn argument would just go away. So, it comes as no surprise that someone will get harmed today somehow, no doubt no sweat. Be alert my bee-hind!

Well, it is now the sixth. We made it through. So I guess things worked out alright. The foreman had it right when he laughed!

The ride up the lift went quick at end of shift today. Now we wait at the station. Whenever the train wants to get here is fine with me. I want to go home. It is dark, 2AM dark. But fairly warm. Sure, we're ten thousand feet up, but it is June. Late spring, sure, but it is nice. Actually it feels warmer out here even at this hour than down in the hole. Not sure it actually is, but it sure feels fine! And the air is fresh, at least for a mining camp. Smelter output, wood and coal stoves, and steam locomotives all make their smelly selves known. There is also a breeze and a hint of flowers and trees and melting mountain snow. Much better than the stale dusty sweaty atmosphere down in the mine. That is the smell of money down there; up here is the smell of life.

And the background noises up here are more gentle. The murmur of a thousand conversations, kids crying, people singing, shouts of celebration. Wagon wheels scraping in the street, horses snickering, the mutter of mine machinery, gusts of wind. It adds up much softer than down in the mine, with blasting, drilling, hammering, steam lines, rocks being thrown into ore carts, men swearing, donkeys braying. Until you come out you don't realize how it hurts the ears.

Getting home and to bed, that's what I want. The wife has made a nice warm home for me and our family. So far it is in a rented drafty cabin. Weather here is not soft, but we have a haven. When me and the wife have more money saved we aim to buy our own house. I hope we can do it a year or two down the road. For now, this spring, maybe if the weather stays fine we can take the children on a picnic. My wages go to make their lives better than the one I had as a youngster. When I get home, I wonder if I dare wake up the wife. She and I haven't been alone in a long time. Maybe I will, just to see her smile and look deep into her eyes. And to get close. Can't wait for that.

The crew I mine with is a good one. We are all on Hamlin's list or we wouldn't be working, not at the Findley or most any other mine in Cripple. I guess that list is alright, at least it lets men like me get a job. We come from many parts of the country, all over the world really. By many different roads we came, and ended up here. We work together and rely on each other as brothers, almost comrades in arms. We get along and the crew has no real complainers, no goldbricks. Good men. I am happy and fortunate to be with such a group. McCoy, Lonklin, Barber, Milheisen, Delano, Kalonski—here we all are, waiting for the train in Independence Station, Cripple Creek Mining District, Teller County, Colorado. It is early in the day or late at night, June six, 1904. God bless us.

I hear the train coming, see its headlamp. Time to form a line so we can get on boar

XIII

THE BLAST WAS SOMETHING TO BEHOLD. ORCHARD LIKED explosions, knew and enjoyed dynamite, would take it over black powder or nitroglycerine any day. He had rigged and arranged countless detonations over the years. Still, this eruption under the station was something different. Maybe it was the target he was destroying, he wasn't sure. The shock walloped him like never.

It eclipsed his first time. It was better than his first kiss, first shot to kill, first blast, first drink, first time with a woman. A grin split his face and he was transported. Owens was for a moment thought fearfully it was a death rictus. To his relief, Orchard's reaction lasted only seconds. Then the reality of what they had done gaped.

The thunderclap rolled. Lumber, pieces of brick, and soft warm pieces of something flew. Debris was still falling as they scrabbled away. Each took care to step on the rags still tied around his shoes. Near a street, they untied the rags. Those were dropped and left by the gutter.

They separated, never to meet again. Owens brazenly walked to his room and crawled under the covers. Sleep eluded him. Orchard climbed on the rented horse and was back at camp soon. Friend and son were sound asleep and never missed him. He wrapped up in a bedroll before his mates rose to start the morning fire. Several days of fishing and relaxing kept him calm and gave him a story. Harry did this for long enough then headed off to Denver and points north. His shadow would not again fall on the streets of Cripple Creek.

BACK AT GROUND ZERO, CHAOS. THE BLAST ECHOED AND REVER-berated over Independence, nearby Goldfield and Victor too. Then, stunned silence.

The engineer was able to stop the train a little way short of the crime. The depot building was still standing but tilted, partly off its foundation. Where moments before men had queued up on the platform was...nothing. Engineer and conductor, the entire crew, ran to the site. Debris was still shifting. Lumber, rocks, body parts were strewn all over. Some men were injured but alive.

There were unearthly creaks and moans. It was hard to tell whether the noise was a man in pain or the slight moving and settling of debris. With this as background, dogs barked, horses screamed, babies wailed. There was a crater where moments before had been a platform. Nearby houses and stores were damaged; windows were broken for blocks around.

As usual miners lined up to board. Twenty seven men heard the whistle blow that morning. For thirteen of them, it was the last sound they heard on this earth. Those unfortunates were physically traumatized. Some were actually blown to pieces. Many more would need amputations or surgery.

The crew of the Findley Mine took the brunt of the carnage. The night crew of the adjacent Shurtloff Mine normally would have been waiting too. Early that June day, by chance they got off shift a few minutes late. The men were in fact running to catch the train. They were on site to help before the echoes stopped. No doubt shaken, they looked for survivors and helped load injured men onto the train to be taken to Victor and medical care.

The scene was Dante-esque. Onlookers and survivors tried to make sense of it. Some towns people started to filter in and helped to lay out the injured. The Sheriff was called as were the owner and manager of the Findley Mine. They came on a special engine with doctors and nurses.

Quickly, word spread. People came. Some only gawked, many helped as they could. Anger and disbelief percolated.

The eastern horizon was graying enough to see. Soon it would be full daylight. Lon was haggard and sick to his stomach as he walked near the ruins.

The 2AM blast was distant but it woke him. Given the warning Ben had given him, he dreaded the noise. Rising, he hurriedly dressed, not bothering to shave or even comb his hair. To the MOA he trotted. The activity there meant there was something big. Lon nodded to the bigwigs he saw, acting like he had work to do there. It paid off; he attached himself to a group heading to the station. There he shoehorned himself on to the train carrying the Findley men and the Sheriff.

He had a nodding acquaintance with Mr. Carlton, owner of the Findley. Lon overheard talk of an explosion, injuries, damage, dead men. What a way to have his fears confirmed! He bitterly wished he had taken more direct action. Silently he beat himself up, finally realizing that just what more he could have done, he didn't know.

The relief train neared the Independence depot, what was left of it. The Sheriff and managers got off and disappeared into the chaos.

Stunned and angry, Lon stood and surveyed the wreck. He had to do something, couldn't stand there and wait. He had already waited too long.

A railroad man pointed up the hill.

"Help us clear the area. There may be men up there. Or...others. Plus, God damn them, whoever set this off, maybe with luck they got injured and we can nab them. Go look. We need to be sure all the injured are taken care of. And find the others." The man held out a pail and gave him a bleak look. "Take this in case you need it."

Up he and others went. They spread out, looking in the gray morning for...evidence, and remains.

Rage propelled him as he walked the hillside above the crater. Already the pail was half full of gore. There on the ground was a hand. He put it in the pail. The first time he picked up part of a miner, he had wiped his hands on the grass. Not now; he was numb. And he was furious.

He saw Ben of all people doing the same.

"Why aren't you at work, Ben, coddling your union pals? Maybe they are planning to bomb the school next, or a womens' luncheon meeting."

"Mines all closed, Lon. As are the bars and clubs. The town is shut down, in shock. We need to pull together."

Lon parroted it. "Together, huh? Let's all go out and gather body parts of innocent men. That will bring us together."

Ben flinched. "Don't you start, Lon. This is inhuman, awful. It is cowardly and wrong. The animal that did this ought to be crucified. No responsible man—or woman—wants anything to do with this. What we have here is not union or owner, it is a horrific crime."

"My God." Lon looked at a lower leg, boot still on. He retched. Then he picked up the poor miner's remains.

Ben picked up his own pail and the two marched their pitiful cargo back to what was left of the station. There a doctor was tending the survivors. He pointed them to an area taken over for a morgue. Bodies were laid out. The pails were put with others containing the earthly remains of several immensely unfortunate men.

The number of people scouring the area was growing. There was some low talk but mostly a stunned quiet. People were staggered that a human being could perpetrate this crime. They were ashamed that anyone in their town would do such a thing. And as the shock wore off, more anger came.

Lon and Ben weren't needed. They were tired and couldn't face handling even one more piece of what had hours before been a man. The two parted ways, going to rest and clean up. The blood washed off easily. Neither could stop the images that came whenever they closed their eyes.

THERE WAS ONCE A VOLCANO OVER CRIPPLE CREEK. AT SOME time in the past it erupted, blew up, formed the giant valley there now. It no doubt spewed so much dust into the atmosphere that the planet's weather was affected for years. Things were never quite the same there after that cataclysm.

That day, history repeated in a way. Enormous anger leading to vengeful rage erupted in town that late spring day. The storm unleashed lasted for many more. Actions and reprisals stemming from the depot blast were stark and dramatic. They forever changed the people and the town.

LON COULDN'T SIT AT HOME. AFTER CLEANING UP HIS ATTEMPT at a nap was backlit by blood red light that wouldn't go away. He went out to walk, hoping to numb his nerves. People were on the street, probably trying to do the same. He saw one whom he knew would have something to say.

"Mr. Hamlin. Isn't it terrible?"

"My God man. Whoever did this...will fry in hell if I have any say. It may take time but we will get him. Or them—it would take more than one man acting alone to do this."

"Could be. Too bad whoever did it didn't get caught in the blast, or falling debris."

"We should be so lucky, Bosini. The sheriff brought in bloodhounds. There was a wire, apparently a tripwire, leading from the explosion back to a building. The dogs couldn't trace any scent. The bombers used kerosene to mask their scent. We found abandoned rags by the street. The dogs are of no use."

Hamlin stopped and wiped his forehead with a handkerchief.

"I understand passions run between owners and union. That is part of it all. But there have been unspoken agreed limits. They have been tested and sometimes crossed, but not like this. This is no angry or passionate outburst. This goes way beyond. It is an intentional and coldly planned murder, mass murder!"

"I understand the mines, saloons and clubs are closed."

"Yes. The town is closed and we are all in shock and mourning."

Lon needed to see his sister. Needed to see if she was alright and how she was taking it.

"Well, sir, if there is anything I can do please call on me. I have to see to my sister now."

"Yes, go to family, by all means. Later, come to my meeting. I have called a meeting today, this afternoon, at three o'clock. We are gathering in the square in Victor across from the Union Hall. We will discuss how to act on this barbaric crime."

"I will be there."

"OH LON. IT IS AWFUL, ISN'T IT? AND BEN IS TAKING IT HARD. He is lying in the back room now. He came, blood on his hands. Miner's blood. Horrible! We got him cleaned up and he is resting, or trying to. He has been tossing restlessly."

"Are you happy now with your union, Abigail? What a bunch of cowardly, blood letting murderers!"

"Lon. This was not union work. They would never do this. Nor approve or allow it. Never."

"Well someone sure as the devil did. Who else but some union loving mad man?"

Ben, scratching his stomach and yawning, appeared. He had laid down but had no more success in sleeping than anyone else. The entire district, most all of the fifty thousand or so, were awake, agitated, upset, fearful.

"It was not a union act, Lon. You and I talked about this earlier. There is no way. The union is for the worker, the man who actually does the work. There is no way the union would agree to killing innocents like this. Even if they weren't members of the union."

"Well, Ben, few will believe that. Big Bill Haywood has too often said he hates anyone who does not support his union. Just how are we to take that? That he wants to buy non union men a cup of tea to discuss things? No, he made his position, and the Federation's position, clear. And you can take this to the bank: those words will haunt him. And they will haunt the people of his organization, the guilty and the innocent alike."

Abby shuddered. "I can't believe you, Lon. You are accusing us of the most vile, barbaric acts."

"Not you, Abby. Nor you, Ben. As I look at this, I have to agree. As extreme as Haywood is, he would not knowingly agree to an act such as this. Not because he disagrees with harming strikebreakers. He'd as soon break their heads as breathe."

Lon took a moment to gather his thoughts and calm down. Even so he repeated himself.

"Haywood wouldn't blow up a station full of men. Not because he loves humanity or because he would hesitate to sacrifice innocent men. He knows it would reflect badly on the union. That's the only reason he wouldn't do something like this."

Lon stopped and took a deep breath. He smiled wickedly.

"This bombing will of course cause Haywood and his WFM to lose support. Support from miners and more importantly, from the community, the towns people. Too bad for Haywood that is likely the work of an independent agent. A mad man to be sure, but it had to be someone acting alone, on his own."

They looked at each other.

Lon went on, his smile returning. "But...The ripples made by that blast will grow far and wide and become like a tidal wave. Much damage will come from this, you watch."

"But you just said it was a lone madman. What damage?"

"Many people didn't want this strike. Folks here just want to go about their lives. They don't want this kind of labor unrest. Heck, even most union miners here in Cripple didn't want this strike. They are pretty much content with their pay and conditions. Now. You watch, this kind of act, this killing, will make the WFM a target. From today on, anything union will attract trouble like iron filings to a magnet."

"But..."

"What can you expect? It sure as the devil wasn't the owners or the merchants who killed these men and destroyed property. It was the union or some crazed person who believes in the union. He made an action, you'll get a reaction."

He looked hard out the window, then turned to them.

"Long story short, be careful. You two had better, Ben especially you, should keep your head down. There will be retribution for this mass murder. And I don't want you to get caught up in it. You too, Abby. Keep your heads down don't say anything, don't go out."

"But we have done nothing wrong." Abby was adamant, angry.

Lon chuckled mirthlessly. "Neither did the miners waiting for the train this morning."

He stared had at Ben. "The poison spewed by Haywood and his lackeys will come back on them. Soon and hard. More people will be hurt. I hope to God no more have to die."

He paused. "Hamlin has called a meeting for this afternoon. A public meeting in Victor. A public gathering today, when people are angry and afraid. I'm not sure this is the time for a meeting, but there it is. Let's hope it doesn't spin out of control—I am concerned that it might. You two, please do not go to that meeting. Stay home, stay here, no matter how much you want to go out and defend your union. Let people cool down before you think about doing that."

XIV

THE THREE LOOKED AT EACH OTHER. WORDLESS, SHOCKED, hurt and angry all described their feelings. The talking was over. Each had their say. They were well aware that their ideas meant nothing to thirteen grieving families, and to many more whose men were injured and in the hospital.

Lon checked the time then carefully put the watch away. He took off his gunbelt and pulled the six shooter out, carefully pointing it down. Then he spun the chamber to confirm it was loaded. He set the firearm down and buckled on the belt with its holster. Making sure the safety was on, he holstered the hogleg. Indecision wracked him. Looking down, his not unkind but firm words bounced off the floor and into Abby and Ben's consciousness.

"It is almost noon. I am going to Victor. To the meeting. I hope to help keep it calm. Now, I can't make you do anything." He looked at Abby, smiling in acknowledgment of that fact.

"But again, I strongly suggest you two stay put. For your own good." He turned and left.

Abby sat silently for about a minute. She stood and went for her jacket and hat.

"What are you doing, Ab?"

"Going to Victor. He can't stop me and I have to see what is going on. And try to keep it calm. The union wants that just like the MOA does."

"Like Lon said, there is anger and gangs are roaming the streets. It isn't real smart to go out now as a known union sympathizer. They could beat me and God knows what they'll do to you."

"Well, Ben, I'm going out as the teacher. I need to know what is happening. Besides, doesn't all that the union stands for mean anything to you? We don't have to wave the union flag, you know. But we have every right to be there. We will just stand and listen. We can hang back, do that, and catch the train back as soon as possible. How can we know what is coming next if we sit here looking at four walls?"

"You're right. How about something to eat first? We'll need our strength and I am finally feeling hungry. We'll do better on a full stomach. Plus, it will give Lon time to get there so he doesn't see us. This has been a long day already, and I suspect it will get longer. Better to go in feeling well and strong."

"Alright. I'll rustle up something. Then we go, right? No stalling."

"Then we go."

Businesses were closed. The streets were fairly empty, quiet, the people still in shock and mourning. Even the few open saloons were not busy. Newspapers were about the only normal thing still out and about. Reporters hurried around talking to all and sundry; extra editions were put out with the latest information.

Word of Hamlin's three pm meeting spread. Many wanted to go. People needed to act, do something positive, not just sit and fret. Plus many wondered what would be said and done. Would there be shouting? Mob action? Or would the gathering be a way to let off steam? Those wanting to know left their homes to find out.

They converged on Victor. By noon a crowd was building. The sun, so enjoyed the day before, felt hot and made folks feel drained and edgy. A jovial gathering it was not. There was a hum of talk and activity which grew in volume as folks arrived. The anger and fear were palpable. People were looking for answers. Couldn't a man even go to work safely anymore? If no answers were to be had, they at least wanted action. The sense of the crowd was, do something. Or find someone to blame and lash out against. Or both.

Lon approached carefully, looking the crowd over. He saw a fellow mine owner who brought him current.

"Hamlin and some others got Sheriff Robertson to resign. You know, the guy who likes the union and sometimes turns a blind eye to their thugs. He is out office. They shook a noose in his face and told him quit or swing. Being a wise man, he quit."

"Oh? Yeah, he sympathizes with the union. Best his sort not try to handle things right now. If he is smart he left town. If I were him I'd be on my way to Denver or Pueblo already. Stepping away was definitely the smart thing. And he is getting off easy—I sure wouldn't want the job of sheriff today. Who is the new man?"

"Ed Bell. General of the militia, at least once he was, not sure if he still is. You know, the old pal of Teddy Roosevelt. Bell's glory was being one of TR's 'Rough Riders.' Charging a hill in Cuba and all. He's also the man who tried to shut down the Victor papers a while back. The time the woman came out with a paper after he marched off with the newspaper men."

"Ah yes, her headline was 'Bloodied but still in the ring' or something like that?"

"That's it. Bell had egg on his face for that, no doubt. But that is past. Today, we need him as sheriff since he is a longtime sworn enemy of the union. There are some militia troops in town too. And I expect we'll see more."

Lon nodded. "Things may get even hotter for a while. I'm glad I thought to bring my little filly."

With this he slapped his holstered Colt .45, grinned, and went on talking. "This is my 'splainer' as I tell my sister— I sometimes use it to 'splain' hard facts to people. Some will listen to no other explanation."

The man eyed Lon's 'filly,' then the crowd, then looked his friend in the eye. "There was a time I would smile at that phrase and your story. But today, it is too damn close to the truth. God knows where things will go. This is bad, Lon, bad."

"I agree. Looks to me like the crowd is chiefly nonunion. I guess there are a few union men. I'd be reluctant to show my face, and ashamed if I had to. But that is just me."

"May be. Hell, you know the union members had nothing to do with that blast. Probably not even the leaders. I imagine they are as appalled as the rest of us."

"You lie down with dogs you get fleas, isn't that the saying?" Unconsciously Lon's hand rested on the handle of his pistol. "If you let Big Bill Haywood spew his venom, sooner or later someone will believe it. No one should be surprised if people act, and react."

The two eyed the crowd silently for two or three minutes. It continued to grow in numbers and noise.

"True, Lon. What concerns me isn't that there are a few union men in the crowd. Hell, I know some of 'em and they are honest honorable guys. I'll tell you something that does concern me. What makes me uneasy, is the site of this meeting."

He looked around, eyes settling on one particular building.

"The Union Hall is just across the street. Sure, this square is a convenient spot, a lot below the Gold Coin shaft house in downtown Victor. This is probably the biggest open space in town. I just hope things stay cool and all we do is talk here. There is room for everyone. Like I said, there are just a few union men in the crowd but you know there are plenty of WFM members in there, in the Hall. Let's hope they don't want trouble. And that the rest of us don't either."

"Yeah. Those union men can listen, should listen. And think long and hard about the direction their so called 'leaders' are taking them."

Lon looked around. He missed seeing Abby and Ben who were standing at the fringes. Shaking his head, he reiterated what his friend had just said.

"I just hope no hothead does anything stupid. Especially because you know many if not most of us are armed. I am, you are. Let's pray everyone keeps their temper, and keeps their hands empty. Like you say, let's hope that today, there's nothing but talking."

"Amen."

The crowd grew. By mid-afternoon the open area was heaving with restless people. They were hot and thirsty, although a few had bottles they took swallows from. People fed off of each other, bored and mad and scared and impatient, upset and looking for answers. Rumors flew and got more outlandish with each retelling.

Abby and Ben lurked, literally. They stood at the edge of the square among a group. Abby saw her brother talking with someone. She lost sight of him as the meeting came to a start.

Right on time, three PM on the dot, thought Lon. A wagon pulled up. There wasn't silence but people paused and watched and the crowd stilled to a murmur. They watched a big ore transfer wagon approach. It was a wagon from one of the big mines. The driver stationed it right at the edge of the crowd, calming the horses. The throng quieted more as Clarence Hamlin climbed up. The man was not tall and even standing up there his presence wasn't particularly impressive.

Lon muttered to his friend, "At least he took the initiative and will address the town. He's no friend of the union but maybe he'll take it easy. I hope he tries to tamp things down and calls for calm. I don't want to think what could happen if he unleashes the dogs."

His friend nodded. "We'll know real quick how the next few days will shape up."

Hamlin cleared his throat, motioned for quiet. After waiting for hours, people took a few seconds to focus on the short man standing on a big wagon. He was about to make up for short stature with big talk. The crowd finally quieted. The secretary of the Mine Owners Association and organizer of the meeting let loose. Lon was appalled. He looked around, hoping to see that at least someone else was too.

"It is time to get your guns and clean up the camp."

Hamlin's opening line silenced the crowd entirely. He didn't stop, didn't soft sell, didn't leave any doubt as to what he wanted.

Many in the crowd would remember the next words to their dying day. Hamlin bellowed.

"Chase these WFM scoundrels out of the district. Make them leave for good and all. Chase them so far that they will never come back. The time has come for every man to take this matter into his own hands."

Most of the folks realized this wasn't a speech. It was war, a declaration of war. In ways it was like the Civil War. Some siblings took opposing sides. Neighbor would shun neighbor. Families would be separated and their property destroyed or lost. Bad feelings would persist for decades. The conflict itself would be short and fought across not a continent but rather within a mining camp.

Lon was about to say so to his friend. Events took over—there was some commotion.

Someone in the crowd shouted a question or maybe it was a jeer. Whatever, the man next to him saw things differently, snarled then took a swing. The fist connected, a fight started. Others started around them, embers on a dry haystack. The flames hungrily licked through the gathering making a bonfire.

Hamlin's unleashed dogs set about.

The fights were bad but really nothing new. Somewhere in Cripple every day, miners were fighting over a mining claim, a girl, a game of pool, or maybe just to fight...Fisticuffs were not a rarity. They usually weren't taken too seriously.

All of a sudden shots rang out. From where—the crowd? The Union Hall? A side street? It happened fast and only the shooter knew.

Horses are sensitive. They mirror the humans around them. If the handler is calm and in charge, the animal remains sedate. If the rider or driver is anxious, nervous and fearful, so the horse. That day, the commotion of the fights and the gunfire on top of the day's tension were too much. The horses wanted to break away, be free of the tumult. With a sudden leap, they bolted and jerked the wagon good. Hamlin was midstream in his tirade, not braced for sudden movement. He was thrown. The driver stopped the horses, too late. Hamlin sprawled on the ground, uninjured but shaken.

It looked like he had been shot off of his perch. Panic. People rushed around trying to escape. The line of fire was to be avoided, if only they knew where it was.

A platoon of militia had been stationed nearby. They were ordered out, double time, to clear the field. The remaining spectators were herded, pushed, forced back to side streets. A few bruises were dealt out and people found they could move faster than they thought. Soon the square was emptied. With the crowd gone, five bodies lay sprawled. Three were hurt and two were dead by gun shot.

Abby stood aside, holding to Ben's elbow. They remained at the edge of the square, in the shade of a building. She looked at the bodies in the sun. One wore a familiar suit.

"My God Ben is that Lon?" She ran to him.

He looked up at the summer sun, listening to shrieks and yells and general noise. His shoulder bled and hurt like a blacksmith was hammering on it. He stirred, for a moment trying to sit. After the first twitch he did his best not to move his arm. Someone was holding him down, helping him to keep still. A familiar voice called to him, but it seemed so far away.

"Oh Lon."

Abby was grateful he was alive. She took a scarf and tied it firmly around his shoulder. "Be still now. Its me, Abby. Ben, help him sit up, no go get a wagon or help and we need to take him to the doctor."

A few people hovered over the other four casualties. Otherwise the square was deserted. Hamlin and his cohorts had vacated the square. The injured were getting help and the dead were carried to the undertaker. This didn't take long, ten minutes give or take.

THE SECOND ACT COMMENCED.

The square was well and truly empty. The town held its breath. And the militia formed up. Again. Orders were to prepare arms and surround the Union Hall. There were men in there, no one knew the situation. How many men? How many weapons? They were quiet for now but their intentions were unclear.

Someone had to find out. Sheriff Bell and Postmaster Danny Sullivan went in. Sullivan, the man who saved T.R.'s life when he was in town as Vice Presidential candidate, was a strong courageous man and a good choice for the job.

They hoped to keep the talking going and stop the shooting. They suspected but did not know if the shots came from there. In any case, the situation needed soothing.

"Men," the Sheriff said, "You WFM members need to come out. Come out unarmed and we can talk. We will not fire on you, you have my word. Men, we have to talk this through. Better to talk than shoot."

"You are a union buster. Your word isn't worth a jar of spit. Not only no, but hell no, Sheriff."

Sullivan: "You have my word as well, men. As a Cripple Creeker, a man and as Postmaster. We need to talk, not shoot. Come on out and let's sit down and talk. You will not be harmed."

"You have our answer."

Bell looked at Sullivan and shrugged.

"This will not end well. Are you sure you won't come out and talk?

"No we will not."

The would-be peacemakers reluctantly left. They felt red hot stares boring into their backs, and half expected to hear and feel the pop of a gun.

Once clear, Bell nodded to the platoon leader.

He yelled the order soldiers dread. "Fire when ready!"

One by one they shot into the Hall. Volley and volley for minutes.

THEY GOT LON TO A DOCTOR.

"This is little more than a scratch, lucky for you. No bone, no major blood vessels hit." The doc rewrapped the shoulder. "You should be alright. Your sister did well with her bandage. I want you to stay here for a while. Go sit over there." He looked at Abby. "I want to watch him for a while. Come back and get him around the dinner hour."

Abby and Ben heard more shots when they stepped out of the office.

As one they headed to the edge of the square to see what was what.

"My God Ben. They're shooting! Americans are shooting at each other!"

Ben nodded, eyes big, face ashen. Abby noticed a smear of blood on his sleeve. Afraid he too had been hit, she grabbed his arm. It was Lon's blood, dry now. His voice was low, gray and tortured and he stared at the gunfight.

"I can't believe it. We are killing one another! This storm has been gathering for months, years. It has broken. God help us."

He pulled his arm away and looked at Abby.

"I heard that George Henderson was one of those killed in the blast. I knew him. He wasn't union but he wasn't a bad man."

He thought of their last conversation, how George said he liked working at the Findley. Ben couldn't take his eyes off the Hall, the sound bullets pinging off the bricks, the yells and smoke.

"What is happening to us? We are devouring each other."

Lon's friend had ducked then fled when the shooting started. Recovered and brushed off, he came over.

"Is Lon alright? He was next to me then ran somewhere then I saw him go down."

"No thanks to you. Yes, he is alright. The doctor is keeping him for a few hours to be sure."

The man sighed.

"That is good. He and I were talking, hoping something like this wouldn't happen. We were afraid there would be retribution, maybe against innocent men."

Ben and Abby nodded mutely. He wasn't done.

"Lon mentioned he had asked you two to stay away. He knew what might happen. Thank God you are not hurt. He knows that, right?"

"Yes, he knows." Abby said this more curtly than she meant to. After all this was a friend of Lon, who had sought her out and asked after him. "Do you think this could get worse?"

The shooting ebbed. A white flag came out of one of the windows.

"Cease fire! Cease fire!"

A sergeant yelled, having several privates run around the corners of the building to be sure everyone got the word. Then, "Down arms!" The uniformed men didn't literally put their weapons down. They did take fingers off triggers and point the barrels downward. The cordon of men stood, watchful and alert.

A squad was ordered in to empty the hall. They went armed, fingers on triggers to defend themselves if need be. Not another shot was fired. The miners were through. They filed out, hands up. All were dazed and fearful and most unhurt. There were four wounded who were promptly taken to the hospital.

Abby couldn't take her eyes off the Union Hall, militia prodding miners, the building pockmarked where bullets hit.

"Ben and I are here despite what Lon said. No one can tell us what to do. But that's not why we came to this 'meeting.' We are here because we too are outraged at the explosion. We want to stand with the innocents, protest the slaughter. Twelve or so this morning is bad enough, and now there are at least two more dead. Ben just told me a friend was among the miners killed. How awful."

The man, Lon's mine owning friend, couldn't stop himself. "Well, Ben, you and probably two or three hundred other men lost friends there. That doesn't make you special."

"No, it just makes me human. Why are we doing this to each other?"

Abby continued. "And, we want to know what happens next. Since you own a mine maybe you have some insight what the MOA plans." Abby pulled Ben to her as if to protect him, more likely to feel protected herself.

By now the Union Hall was empty of people. The surrendered miners were marching away under armed guard. Stretcher bearers had taken away the wounded. She idly wondered what would be done with the miners. Abruptly her wonder about what next was answered.

Of a sudden she saw groups filter into the square. The area was not full but there were twenty five or thirty men. By dress and action they looked like either miners or hired security guards. She couldn't tell. There were no firearms evident but a few had clubs or iron bars.

These idlers stood around then spontaneously formed into gangs of six or eight each. The Hall was stormed in a rush. It was almost as if they expected opposition. She heard yelling and saw arms waving. The invaders acted angry and clearly intended no good.

Sounds soon came out. To her dulled senses it was a background rumble, almost distant thunder. But it became a shrieking symphony with several movements: glass breaking, plaster walls being knocked apart. Another part was furniture being smashed and pieces thrown out the glassless window openings. To add visuals, pages of ledgers books and documents, torn and shredded, floated in the breeze. It didn't take long. The Hall was soon destroyed, a hulk with gaping, sorrowful holes and bullet scars.

For some crazy reason it looked like photos she had seen of the Alamo, old, bullet worn and forlorn.

It was unbelievable. Not sixty minutes before in the square men and women stood, peaceful if not friendly, to listen to Clarence Hamlin. Now the crowd was scattered, the ground was bloody, the Union Hall was destroyed, its members marched off by the militia.

Worse, more small angry groups were forming. They too looked like hired security or miners. No doubt liquor was available even if the saloons were closed. Buzzing and swarming like African bees, they worked their ways through town. Abby wondered but somehow knew what they would do. Every union owned building, business, and member would be a target for these packs.

Lon's friend took the reins. "You two had better get out of here. You need to go. No one with any tie to the Union is safe now, not here. Come, I rode the train over but have a carriage on hold. We'll take that back. I will drop you at Lon's home. You should be safe there." He offered his arm to Abby and they walked off as if strolling in a park. Ben followed.

Later, Ben and Abby sat quietly, trying to take it all in. The door opened and in walked Lon. He seemed to feel well enough. His left arm was in a sling but there was no bleeding.

"Lon!"

"Hello you two. It is very good to see you! I thought I was a goner there for a few minutes. I do remember your tying something on my shoulder, Ab. Thank you."

"We're just glad you are here and able to move around. And you can thank Ben, he got help to you quickly."

Lon held out his good hand. "Thanks, Ben." They shook.

Ben joked feebly. "And here I thought mine owners bled money not red. That was a surprise!"

"The whole town is bleeding tonight. The innocent miners, the people at the meeting, and no doubt there is other blood being shed tonight. Feelings are running high."

Abby sat and flounced.

"I can't believe it. Shooting. American troops firing on American citizens. And citizens firing back. My God, thirteen dead at the station, two in Victor, two at the Vindicator, and who knows what else. I hope no one else is hurt."

Worked up as she was, emotion almost overcame her. She choked a sob.

"And there is looting. Packs of angry men taking to the streets doing what they want. Some taking out anger on the union, others using the opportunity to get even for past slights and disagreements. I have never in my life felt so afraid and deserted. I am so glad you two are here and are safe."

"The man or men who set off the blast went too far. But what the owners are doing is bad too."

Lon growled. "The owners aren't roaming the streets. Men are damned tired of the violence and intimidation and they are giving it back to the union and its men. Is it right? No. But it is understandable. There was absolutely no call for the explosion killing a dozen or so and maiming many more. If you do that, or associate with someone who does that, you damn well better not whine when you get the same back."

Ben made a 'who knows' gesture with his hands. "You are correct, Lon, this mob rule is not right. Maybe it is a long held back reaction to the union, I don't know. And yes, I do have to say, I felt vulnerable out there after we got you taken care of, Lon. There's no telling what might have happened if a group got hold of us, me and Ab. We owe your friend more than we can ever repay. He got us back here in his coach. We thanked him and next time you see him, please thank him for us again."

Lon nodded. "I will do so."

He stood, shuffled to a closet, pulled out a cane and went to the door.

"I am going out. Get the lay of the land, so to speak. See who is doing what to whom and all."

Abby smiled weakly as the door opened. "Be careful."

"Oh, no one will bother an old man with a cane and an arm in a sling. I'll be back soon as I can. Sit tight!"

He was back soon, eyes big and face grey.

"That was quick. So, Lon, what is going on out there?"

"Not good, not good at all, Abby. It is worse than we thought. I was a little concerned for myself out there, and like I said, I'm an old guy with a cane. Time to stay in, lay low, not show your face. Especially you, Ben."

"Why? I've done nothing wrong."

"It doesn't matter, man! You're known as a union man. Like it or not, that is your reputation. And things are...out of hand. The intimidation and beatings the union have been giving out have finally come home to roost. Not to mention the dynamite killings."

"I had nothing to do with..."

"You're not listening. It doesn't matter that you had nothing to do with it. Gangs of men are roaming. Predatory gangs are in the streets. You saw the Union Hall there in Victor? Every hall around has been ransacked like that. And union owned stores. And known WFM men—no women yet..." With this he looked meaningfully at Abby. "No women yet, but some of the WFM men are being beaten. If not beaten, rounded up and imprisoned."

"This is practically civil war."

"Yes, it is. And it could spread. If you want to stoke the flames, go out and try your luck. Seems to me the best thing to do is lay low. Let the anger and venom come out so people can cool off."

"Sounds like the easy way out." Ben was torn. He wanted to be sure Abby was alright but he felt the need to stand up for the union.

Lon had some venom of his own he had to vent. "And dammit, I told you so. The people and the mine owners are done with Big Bill Haywood and his hateful violent organization. You are seeing the death throes of the Western Federation of Miners. And none too soon, but too late for thirteen unfortunate miners. And others."

"We can't stay here. We need to go to our homes."

"Sooner or later, yes, you will have to go back. But not yet. Abby, Ben, please. Tonight, stay here. There will be more unrest tomorrow. Lay low here. For the night. Wait and see."

NEXT MORNING, BEN AND ABBY WENT TO THEIR HOMES EARLY and quietly. The mobs had played out, sleeping off the anger and the liquor. The camp was coming back to life. Mines were open again. Men went to work. Stores, schools, and the rest of daily life resumed. People were as tense as a politician in a recall election. Still, day to day routines were being followed. The towns were functioning. Maybe not at full tilt, but they were functioning.

LON WAS CALLED TO ANOTHER MEETING. HE FELT WELL enough to go on his own. This one was small and discreet. The public knew nothing of it.

Sheriff Bell presided. He looked the room over. Hamlin and a few of the big mine owners were in the back watching to see that things got done. In front were a handful of every day Cripple Creekers and Victorites. They sat. Lon went to the front, not hobnobbing with the heavy hitters.

Bell glanced at the men standing in back then spoke.

"You seven are solid citizens of the District. You are merchants, small owners, and the like. We need your help. Things came to a halt yesterday and we need to get things moving once more. Normal life has started up again and we need to help it along."

"What can we do that you and the militia can't?" Lon had the same thought but a hardware store man asked it first.

"The WFM needs to be combed out of this camp. We need you to examine, talk to, union members and decide who can stick around and who has to go. Those who are not ne'er do wells or trouble makers can stay here and resume their lives and keep their jobs. But first they have to renounce the union in writing. Those who will not disown the union or who are undesirable will be deported."

"What is an undesirable person or behavior?"

"That, gentlemen, is up to you. You will decide what makes a man undesirable in Cripple Creek."

The seven men were looking at each other, appraising and wondering.

A man Lon knew only by reputation reacted. "So you want us to try these people according to our whims?"

"No. Whatever you decide needs to be evenly applied to all. This is not an opportunity to even an old score. It is a chance to rid the camp of the union."

Bell paused, looking at each man.

"You know as well as do I who the troublemakers are. They aren't the man who goes to the saloon after work or who argues with his neighbor. They are the men who interfere with normal lawful business. If you want it written out I will do it, but the guidelines will simply repeat what I just said. This isn't the time for fancy rules and procedures and niceties. We have to act fast."

The man nodded, not happy with the answer. "And you say they can't stay. Where do they go?"

"We deport them. Get them out of town, clear out of Colorado."

Lon thought this too much. "Deport them out of state? How, where?"

Bell's smile was wintery and flinty.

"Arrangements have been made. We will put them on a train, under guard if need be. The train will take them to the Colorado border. New Mexico or Kansas can have them. If they have families, the wife and children can follow later. We don't care. That sounds cold, but what happened here yesterday makes deportation look downright merciful. The thing is, we need to get rid of the troublemakers before they blow up the whole town. And, I have to say, before a lynch mob takes after them. We want to, have to, cleanse this mining district of the WFM."

"Sheriff," Lon asked, "What is the standing for this? By what legal theory are we acting? If all a man has done is join a union no crime has been committed. Is deporting them legal? Aren't we asking for trouble on this?"

"I'm glad you asked that, Bosini. We need to be clear that this is perfectly legitimate. The law allows for dispersal of a mob. The WFM is a mob! Yes, this action is appropriate and within legal bounds." He again looked the citizens in the eye, one by one, and continued.

"The work needs to start this morning. We have lists and will send the people to you. Ask them anything you want. If there is any doubt of their loyalty or reliability, get rid of them. Trains will be waiting. Are there questions before you begin?"

It was clear what was wanted; no one spoke. The sheriff nodded and left the room with the big shots.

Some of the men begged off the morning meeting. There were only four men left as Lon and two others cited loose ends they had to tie up. Leaving the room, they went their ways. Lon headed home, making a quick stop at the telegraph office. He was glad to find the only one message needing sent before his.

BEN WAS UP EARLY ANYWAY AND HE KNOCKED AROUND THE place, nervous energy and indecision making him pace. After thinking it over, he opted not to go to work. He had to agree with Lon, it made sense to lay low. Hopefully things would calm down. Hungry to know what was happening, he needed to see a paper. He snuck out and stopped a newsboy selling papers.

The headline said it all: DEPORTATION!!

He saw a man from his crew who had also decided to skip the day.

"Hello McNall. Bad times, no? I heard that about twenty five WFM members got sent out yesterday."

"Sent out?"

"Yup. Most were in bullpens already, some got rousted from their homes. They got put on a train, and, bye bye. I guess that their families and all are still here. Those men were loaded aboard, at bayonet point, and were kept under armed guard. And the train headed south. Rumors are they'll be let off in New Mexico. God knows what reception awaits them there."

Ben was surprised but in a way he was not. After all, the explosion was taken for a WFM declaration of war. He wondered if the yayhoo who did it realized the repercussions. Whoever would do such a thing probably didn't care, truth be known. The rank and file WFM members would pay the price for the criminal's two boxes of dynamite.

Ben shook himself back to the here and now. "That is not good news. The owners are going to stomp the Federation to death over this. I look for more miners to be sent away in the next few days."

The man looked appraisingly at Ben. "You're a member aren't you? Aren't you and Big Bill Haywood pals? You can be sure that they'll come looking for you. Better be ready, or get out of town on your own. Where is that big ape anyway?"

Like most Cripple Creekers Ben was angry. Many were angry at the WFM generally. Ben was furious at Big Bill Haywood. His bitter tone surprised even him.

"Oh I'm sure he is safe and sound. He's probably drinking champagne, holed up in Denver at the Union offices. He sure as hell isn't here to face the firestorm. Maybe he personally didn't set that blast off, in fact I'm sure he doesn't have the balls to do it. But he sure piled up the brush and kindling so the fire would burn big and hot. No doubt he is happy to

260 | STAN MOORE

let us front line workers take the heat. He will never return to Cripple Creek."

A messenger came by with a telegram. "Are you Mr. McNall? Ben McNall?"

He warily nodded. "That's me."

"Sign here for your telegram please."

Ben opened the envelope and read. 'Go Abby's house. Utmost speed. LB'

"I had better see to this. Good luck, hope to see you on shift tomorrow." He turned and left.

ABBY WAS NOWHERE TO BE SEEN. THERE WAS SOMEONE IN there. Through the window, Ben could see a man pacing in the front room. Adrenaline surged. Who is in there? Is Abby alright? He threw open the door, hand on his pistol, prepared for anything—a fight, bad news, something.

Lon stopped and looked up.

"Ben. You made it."

"Where's Abby? Is she hurt? Is she alright?"

"She's fine. Not hurt. In fact she is in front of her class at school. Trying to get her life back to normal."

"Why..."

"Ben, the owners are deporting anyone in or around the Union. If a man agrees to renounce the union he can stay. Otherwise, he is not welcome here. Will not find a job. Might find himself in an alley with some explaining to do."

"So I heard. Is this America or some tinpot country where the king can order someone's head off?"

MISTER HAMLIN'S LIST | 261

"A king, no. A sheriff, I am afraid, yes he can. Not order a head cut off but he can and will order a man out of the state, yes."

Ben sat. "Wonderful. What do we do now? Won't the Governor intervene and give us due justice? Those aren't the words, it is due process. Innocent until proven guilty and all that. Any possibility of that happening?"

"Right now, the Governor is standing back, as are the Attorney General and the Legislature. They have said this is a local problem and is ours to handle as we see fit. The owners—we—have installed a new sheriff and he is taking steps. There is a committee, a citizens' committee. It will hear the cases of Union men, one by one, and decide if he stays or goes. Sheriff Bell wants to calm matters and root out the problem so we don't have more Americans killing Americans. And Ben, I am on that committee."

"That blast wasn't the Union!"

"So you say, Ben. But, have the mine owners or anyone else gone out and killed people they disagree with? People who have done no wrong, have done nothing but try to do their job?"

Ben stared at the floor.

"So you are telling me, Lon that the price for me to stay here is to turn my back on the Union."

"It looks that way. Now, understand, Ben, I will try to shield you. But you aren't just a miner swinging a single jack. It is well known that you have Big Bill Haywood's ear. You have a target on your back, my friend. Think about it: Would it be so horrible for you to distance yourself from him and his thugs? And think about Abby. Do you want to leave her adrift? Thank God I'm here, scant satisfaction that will be for her. She wants you, not me. You know that the miners being deported are having to leave their families here, at least for now."

Ben continued to stare at the floor, but he was really looking inward, weighing and agonizing.

"Ben, you would be a real catch for one of the gangs sweeping the streets. Every one of them would love to 'persuade' you to get on an outbound train. Come with me, talk to the committee this morning. We can smooth this over. You and Abby can go on with your lives."

"I don't know, Lon. I just don't know. I have to go and think on this." He went to the door.

"Ben. At least stay here to think it over. Out of sight."

He nodded, came back in and closed the door.

They traded places. Lon went to the door, Ben sat.

As he was leaving, Lon murmured. "I'll come back when I can. Think long and hard, Ben. This is a watershed day. For you, for Abby, for everyone in the camp."

UNPLEASANT TIMES ROILED ON. THE CRIPPLE CREEK MINING district writhed and hurt.

The militia kept a rough peace. They did not allow lynchings or arson. They did look away if a man got dragged out of his house or a business. Groups of civilians patrolled streets throughout the district. They were looking for union men. When they passed a union store or hall they made sure the previous days' vandalism was taken to another level. Anyone they found who was known or suspected to like the

WFM was dealt with. None were killed. Some got to go before the committee to plead their case. Some were taken directly and put aboard one of the waiting trains.

The town was open for business again. Stores, schools, clubs, and other establishments did their best to go about their daily routines. People were still upset and edgy in general. At school, Abby found students too excited and unfocused to learn much. It was a long day for her and them. She looked forward to relaxing at home.

"Ben! Why are you here at my house? Didn't you work today?"

"Lon met me here. He called me here. Said it was because he thought it a safe place,"

"Lon? Why a safe place?"

"Abby, the owners are using that depot blast as an excuse to erase the Union, stomp it to death. They are deporting Federation members. At least members who won't renounce the WFM. There's a committee—Lon's on it—which is 'interviewing' miners to decide their future."

"Interviewing?"

"Yes. The committee is made up of seven Cripple Creekers—all non union of course. Men are being swept up by gangs and brought before that group, that self-appointed committee. The committee men can ask any questions they want, of course about union and union activity. If the man disavows the union, good for him. He can stay with his family and work. If not, well, good bye. He is deported. Sent away. This is real Star Chamber stuff."

Abby had the stray surprise that Ben knew about the Star Chamber, Medieval England's summary committee with life and death power. Then the word 'deported' registered.

"Deported? They're deporting Americans?"

"Yup. Loading them on a train like cattle. No court order, no hearing, no lawyers. They are just loading miners up and sending them to another state. Exactly how, I don't know—if they're let off just over the border or sent to a city or what. No matter how it is done at the far end, what kind of reception can the exiled miner expect? He has no money, only the clothes on his back. What is a miner from up here going to do in Kansas? This is a bad business."

"And you, Ben. What will you do?"

"Well, whatever it is, I want to make the decision myself. If I decide to stay with the Union, I'll go down to the station and get on board myself. If I decide to keep mining here and drop the Union, that's what I'll do. What I don't want is to be herded up like a damned longhorn steer and forced onto a train to nowhere."

"So. What will you do? I don't want you to go, Ben. If you decide to take that train ride, you are finished mining. At least in Colorado, probably across the west. Your name will be on a list and no one will hire you. And what about us?"

"Well, if it is alright with you I will stay in here for a few more hours. Give me time and quiet to think. Right now I'd just as soon stay away from the gangs."

She thought a moment, looking hard at him.

"Stay and think, fine, Ben. Do what you will. You need to know that I will stay here in Cripple. I am a teacher and my students depend on me. If you don't care about us, I will not run off and follow you to God knows where."

"Abby..."

She ignored him. "Just remember, before you decide, that most miners here are happy with their pay and conditions.

Most did not want to strike. Even when the WFM tried to stir things up in Idaho Springs, Telluride, and other camps the miners really never bought it. This whole thing has really been just a power play between Haywood and the MOA. An evil, killer power play. I hope those WFM and MOA men playing childish games with our lives are happy."

"So tell me, Ben McNall, do you really want to get caught up in that, give up your life, our life, for that?"

"I don't know what I want, Abby. I want you but I don't want to turn my back on the Union."

"I was a supporter of the Union. I believe people should be able to get together and negotiate their work conditions. But when they started killing innocents, they lost me. Tell me, Ben, what has the Union done for you, other than make you hide, a fugitive?"

"I don't know, Abby. I do not know."

"And what can you gain, Ben, by leaving me and Cripple with a bunch of other refugees? Going to nowhere, with no job, no money, no future?"

XV

Lon, representing the committee of seven, met with the Sheriff and Clarence Hamlin. The busy, exhausting time had gone fast and they were reviewing. Bell spoke first.

"Let me summarize: It has been seven days since the Independence Depot blast. We have made good progress. The weasels at the Western Federation of Miners have been rousted out of their holes and sent away. Haywood squawks in Denver but he won't be back. Those miners who saw sense about the union have stayed and are doing good work. We should have done this back in '94."

Bell's week had been fuller than he ever wanted to experience again. The Sheriff as usual had to deal with bar brawls, domestic disputes, petty thievery and other human behavior usual to a camp of fifty thousand. On top of that, he had to investigate the blast, round up WFM members for interrogation, hold and deport those deemed undesirable, and coordinate with the Militia.

"Whoever set off that explosion planned it out well. We brought in hounds to trace them. We had at least one man's scent from the pull handle, and saw where he or they laid up, hopefully leaving something. No luck, though. The dogs picked up the scent there but lost it pretty quick. We found kerosene soaked rags which the man or men tied to their shoes in order to cover their tracks. In the dark and confusion they got back to town and got away. So we don't know who or where they may be."

Clarence Hamlin nodded. He too had been busy.

"Yes, my staff and I have been hard at it as well. The committee has passed many, probably most of the miners, to work. So there are plenty of work permits to fill out and record for those men. My list grows.

Lon looked over at the two architects of the mine owners' response to the blast. He thought of the effort to get rid of the WFM men as a purge. It was almost like in Russia with the pogroms he read about. Where the Tsar's men went house to house, forcing people of a certain religion to leave. The troops didn't care where the people went as long as it was away. Whatever the similarities, Lon was careful not to say the word out loud. Like most Cripple Creekers he was watching his Ps and Qs right now.

He did ask a safe question. "Well, things are back to more or less normal. The mines are open as are the stores, clubs, and saloons. How many miners have been deported?"

"Over two hundred. We have sent ten score of the WFM's finest to Kansas and New Mexico. Actually our trains stop just short of the border. Our transport doesn't cross the line. The ex-miners have to get out and walk."

"Oh? What do they do?"

"Try to find work I imagine. Some will drift east but some will come back. Not to Cripple probably but to the mines. That is why we keep lists, and will be sharing them with owners in other mining camps."

"Whoever did that blast sure muddied things up for a lot of folks, didn't they? Miners and their families thrown out of camp and having to move and try to find work. What the heck was the thinking with that explosion? Trying to scare miners into joining their ragtag union? That sure backfired!"

Hamlin went on.

"Yes it did. Getting back to your question about the deportations. I guess a sheriff in western Kansas met one of our gift loads. He turned the ex miners back, wouldn't let them come in to his state. Flat out refused them entrance. So they walked west, back to the town of Holly. I guess they were able to spend the night there. So I understand."

"Tough going for them."

"They should have thought of that before they allowed the Union to start blowing things up. It all comes back to their letting Big Bill Haywood take control. If the rank and file had their way, likely none of this would have happened."

Bell chimed in.

"Frankly I don't care what happens to them. We have names and have sent the lists out. The WFM is done for. There are damn few Colorado mines which would hire their members now. Probably damn few mines anywhere."

Lon frowned.

"Are any men left in the 'bull pens'?" Bull pens were open air prisons. The militia and owners set up fenced areas near the station. Suspected Union men were held there until they either quit the Union or were loaded on trains. He had read of similar open air jails used by the British in the Boer war, to round up enemies and keep tabs. He shoved that thought back and away.

"Or are they empty and all the troublemakers gone?"

Bell shrugged. "The pens are nearly empty. Most cases have been resolved. Men either stayed with that damned Union, and a few did, why I don't know. Or they gave it up, renounced it. And then they went back to work, the sensible thing to do."

Hamlin smiled, glad to be on top of the job at hand.

"Either way the bull pens are about ready to be closed up. They are—were—an ugly but necessary part of our recent history. Just a few men have yet to be dispensed with."

He too shrugged. "Well, now that the MOA has put paid to those socialist bombers, we can get down to the mining business. There is gold to be dug! I have a meeting so must go. I will talk with you tomorrow or the next day, Bosini."

LON OBSERVED A BYPRODUCT OF THE UNREST. "PEOPLE ARE moving out of the district. Not just the families of the deported, following them. Others have left or intend to. I've been told that many think the potential riches are not worth the danger."

Bell shrugged. "Good riddance."

ABBY DIDN'T GO TEACH THIS DAY. SHE GOT UP, DRESSED, AND sat at the breakfast table. She wasn't hungry and didn't even try to eat. For a while she sat with her head in hands, eyes closed. Finally she rose and left the house, going down to the main street.

There were just a few men in the bull pen. She walked in fairly near the barbed wire. A militia man officiously nattered at her. He actually unshouldered his rifle, as if a lone woman was a threat. "Close enough, ma'am. Stop there."

She ignored him and walked right up to the wire. She put her hand on the top rail of the enclosing fence. The militia man continued to hold but didn't aim or ready his rifle. He did watch her anxiously. Regardless of orders he was reluctant to stop or lay a hand on a lady. Still, his orders said nothing could be handed across into the bullpen.

From inside, Ben approached, hands in his pockets. "Hello Abby. Thanks for coming down." He looked over his shoulder. An engine with just one box car chuffed into the station. He knew.

"I think that train is for us, this last group of us no good union men. Cripple will at last be cleansed."

"It is not too late, Ben. You can still resign from the Federation, get a job or run the Double I. Please, McNall, do it, if not for you, for me, for us."

He looked deep into her eyes, slowly nodded a sad no, turned and walked toward the cattle car.

Afterword

THE LAST HALF OF 1904 SAW CHANGES IN CRIPPLE CREEK. Many people left. Of course the families of the deported miners followed their menfolk. Others deplored the violence and tumult. Wealth from the big mines peaked about then. Overall employment dropped. For many reasons, population in 1905 fell to about 28,000.

The Western Federation of Miners never recovered. Cripple Creek mine owners were able to drive membership down. Similar efforts in other mining camps sent the WFM into a death spiral.

Big Bill Haywood left the WFM and went east. He was active in the Industrial Workers of the World, the 'Wobblies,' in 1914. This was a radical, violent industrial union. He fled to the Soviet Union where he was given hero status. He died there in 1928.

There are reports that folks living near the Independence Station were warned to vacate on the night of June 5. Neither a source nor credible evidence of this were found by the author.

Harry Orchard's real name was Albert Horsley. He successfully got out of Colorado in June of '04 and returned to

Idaho. There he assassinated ex-Governor Steunenberg. His weapon of choice was dynamite in the mailbox, rigged to go off when opened. Orchard did this to avenge his loss of wealth when the Governor forced him to flee Idaho in the late 1890's. The man was not charged or prosecuted for either the Vindicator or the Independence Station explosions. He was imprisoned for the Steunenberg murder and died behind bars in 1954.

Clarence Hamlin, Spencer Penrose, Charles Tutt and other members of the MOA went on to live wealthy, long lives in Cripple Creek and Colorado Springs.

George Henderson is the author's great uncle. He rests in the family plot in Colorado Springs.

About the Author

STAN MOORE IS A HUSBAND, FATHER, GRANDFATHER; A THIRD generation Coloradan; an author and historian; a Vietnam veteran; a retired small business owner; and an avid mountaineer, backpacker and desert rat. Moore and his wife make their home near Denver with two cats who let them stay there.